SUPER SOLDIERS

Emerging Technologies, Ethics and International Affairs

Series editors:

Jai C. Galliott, The University of New South Wales, Australia
Avery Plaw, University of Massachusetts, USA
Katina Michael, University of Wollongong, Australia

This series examines the crucial ethical, legal and public policy questions arising from or exacerbated by the design, development and eventual adoption of new technologies across all related fields, from education and engineering to medicine and military affairs.

The books revolve around two key themes:

- Moral issues in research, engineering and design.
- Ethical, legal and political/policy issues in the use and regulation of Technology.

This series encourages submission of cutting-edge research monographs and edited collections with a particular focus on forward-looking ideas concerning innovative or as yet undeveloped technologies. Whilst there is an expectation that authors will be well grounded in philosophy, law or political science, consideration will be given to future-orientated works that cross these disciplinary boundaries. The interdisciplinary nature of the series editorial team offers the best possible examination of works that address the 'ethical, legal and social' implications of emerging technologies.

Forthcoming titles:

Culture and Human-Robot Interaction in Militarized Spaces
A War Story
Julie Carpenter

Social Robots
Boundaries, Potential, Challenges
Edited by Marco Nørskov

Super Soldiers

The Ethical, Legal and Social Implications

Edited by

JAI GALLIOTT
The University of New South Wales

MIANNA LOTZ
Macquarie University, Australia

LONDON AND NEW YORK

First published 2015 by Ashgate Publishing

2 Park Square, Milton Park, Abingdon, Oxfordshire OX14 4RN
711 Third Avenue, New York, NY 10017

Routledge is an imprint of the Taylor & Francis Group, an informa business

First issued in paperback 2017

British Library Cataloguing in Publication Data
A catalogue record for this book is available from the British Library

The Library of Congress has cataloged the printed edition as follows:
Galliott, Jai, author.
  Super soldiers : the ethical, legal and social implications / by Jai Galliott and Mianna Lotz.
      pages cm. -- (Emerging technologies, ethics and international affairs)
  Includes bibliographical references and index.
  ISBN 978-1-4724-3295-7 (hardback)
  1. Military art and science--Technological innovations--Moral and ethical aspects.
  2. Military art and science--Technological innovations--Social aspects.
  3. Soldiers--Miscellanea. 4. Medicine, Military--Miscellanea. 5. Military robots--Moral and ethical aspects. 6. Robotics--Military applications--Technological innovations--Moral and ethical aspects. 7. Robotics--Military applications--Technological innovations--Social aspects. I. Lotz, Mianna, author. II. Title.
  U42.5.G35 2015
  174'.9355--dc23
                                                                    2015022846

ISBN 978-1-4724-3295-7 (hbk)
ISBN 978-1-138-57652-0 (pbk)

Contents

Notes on Contributors

Jai Galliott is a Post-Doctoral Research Fellow at the University of New South Wales in Sydney, Australia. His research interests revolve around emerging military technologies, including autonomous systems, soldier enhancements and cyber warfare. His secondary interests include military strategy and applied ethics. He holds a PhD in military ethics from Macquarie University and as a former Naval Officer in the Royal Australian Navy, now conducts contract research for the Department of Defence. His recent books include *Military Robots: Mapping the Moral Landscape* (Ashgate 2015) and *Ethics and the Future of Spying* (Routledge 2015).

Barbara Gurgel recently graduated from the University of Massachusetts Dartmouth. Her career started with an internship during her last semester at the University, editing strike details for the UMass Drone database. Barbara is currently working with Professor Avery Plaw on a project about the Armed Forces' use of education in the creation of better soldiers. Barbara's areas of interest include international relations, foreign languages and extraterritorial uses of force.

Katrina Hutchinson is a Post-Doctoral Researcher at Macquarie University. Her current research focus is on the ethics of innovative surgery and surgical research. She previously completed a PhD on free will and retains a research interest in issues of human agency and moral responsibility. She has a further interest in feminist philosophy and has co-edited a book on women in philosophy which was published by Oxford University Press in 2013. She is also interested in questions about the goals of philosophy and the role of philosophy outside of academia. She maintains an interest in higher education teaching although her current position is research only.

Armin Krishnan is currently Assistant Professor for Security Studies at East Carolina University (USA) and has previously taught intelligence studies at the University of Texas at El Paso. He received his PhD in 2006 from the University of Salford, UK, and also holds a Masters degree in Intelligence and National Security from the University of Salford (2004) and another Masters degree in Political Science, Sociology and Philosophy from the University of Munich (LMU). He specialises in defence and intelligence and has published altogether three books on aspects of contemporary warfare. His most recent book was published in German language by Matthes & Seitz Berlin Publishing in 2012 and dealt with the topic of targeted killing as a tactic and strategy of war.

Alex Leveringhaus is a Post-Doctoral Researcher on the research project 'Military Enhancement: Design for Responsibility and Combat Systems', funded by the Dutch Research Council (NWO). This collaborative project between the 3TU Centre for Ethics and Technology and ELAC is based in the 3TU Centre at The Delft University of Technology, Netherlands. Alex is based in ELAC at the University of Oxford, and works in collaboration with fellow Post-Doctoral Researcher, Tjerk de Greef in Delft. Alex holds a PhD in Government and a Masters in Political Theory from the London School of Economics (LSE), where he also served as an LSE Fellow in Political Theory from 2008–10.

Mianna Lotz is Senior Lecturer in the Department of Philosophy at Macquarie University. She has published work in the fields of bioethics, applied ethics and social philosophy, with a particular focus on questions of parental liberty, the family, adoption and children's rights and interests. Her work has appeared in a variety of edited volumes and journals, including *Bioethics*, *The Journal of Social Philosophy* and the *Journal of Applied Philosophy*.

Steve Matthews is Senior Research Fellow at The Plunkett Centre for Ethics at St Vincent's Hospital in Sydney. Steve has worked on a range of projects related to questions of personhood and ethics. His published works relate to the metaphysics of personal identity over time, the philosophy of psychiatry and applied ethics. Recently he has worked on questions of privacy and anonymity, particularly in the context of applied philosophical questions about the internet. More recently he has been focusing on questions of autonomy, agency and narrative identity where those concepts may test, and be tested by, empirical findings related to those struggling with addictions, particularly in the drug and alcohol field. This latter work is in conjunction with a team of researchers funded through a large Australian Research Council grant.

Seumas Miller is Professor of Philosophy at Charles Sturt University and Senior Research Fellow, 3TU Centre for Ethics and Technology, Delft University of Technology. He was Head of the School of Humanities and Social Sciences at Charles Sturt University 1994–99 and Foundation Director of the Centre for Applied Philosophy and Public Ethics: An Australian Research Council funded Special Research Centre (2000–07). His extensive publications include writings on social action and institutions, terrorism, business ethics and police ethics. He has also been awarded numerous competitive grants and consultancies.

Avery Plaw specialises in political theory and international relations, with a particular focus on strategic studies. Before arriving at the University of Massachusetts Dartmouth he received a two-year fellowship from the Social Science and Humanities Research Council of Canada (SSHRCC) to pursue research at New York University. He published a book in 2008 entitled *Targeting Terrorists: A License to Kill?* in Ashgate's Ethics and Global Politics series and he is currently co-editing a collected volume entitled the *Metamorphosis of War* (Oxford: Inter-Disciplinary Press). He has also edited a collected volume concerned with managing the challenges of philosophical and cultural pluralism entitled *Frontiers of Diversity: Explorations in Contemporary Pluralism* (2005). Prof. Plaw has published numerous peer-reviewed articles and chapters, and in 2008 he won the American Political Science Association's Wilson Carey McWilliams prize. His book *Targeting Terrorists* was shortlisted (with two other books) for Best Book of the Year in International Relations by the Canadian Political Science Association.

Joseph Pugliese is Research Director of the Department of Media, Music, Communication and Cultural Studies at Macquarie University. Joseph's research and teaching are principally orientated by issues of social justice. He deploys critical and cultural theories in order to examine and address the relationship between knowledge and power, issues concerned with discrimination and injustice, state violence, institutional racism and regimes of colonialism and empire. He examines these issues in the context of everyday cultural practices, the state, institutions of power such as law and the interface of bodies and technologies.

Wendy Rogers is Professor of Clinical Ethics in the Department of Philosophy and Australian School of Advanced Medicine at Macquarie University. Her research interests include public health ethics, research ethics and the ethics of evidence-based medicine and surgery. She has published in a wide range of journals including *Bioethics*, *Journal of Medical Ethics*, *International Journal of Feminist Approaches to Bioethics*, *Lancet*, *Mayo Clinic Proceedings* and the *American Journal of Public Health*. She is co-author of *Practical Ethics for General Practices* (Oxford University Press 2004 and 2008).

Joseph Savirimuthu is Senior Lecturer in Law at the University of Liverpool. His research principally involves analysing regulatory challenges and issues posed by new and emerging communication technologies for traditional approaches to governance. Some of the areas examined include issues such as surveillance, identity theft, child online safety, peer-to-peer file-sharing controversies, online dispute resolution and managing personal and corporate identities. Joseph is now exploring the legal, ethical, social and technological challenges posed by autonomous systems and robotics as they relate to aging, healthcare and warfare.

Robert Simpson is a Lecturer at Monash University. His research interests are primarily in social and political philosophy and include: free speech, hate speech, the analysis of speech-harm, attributions of responsibility in law, the moral limits of the criminal law, the ethics of human enhancement, the epistemology of disagreement and philosophical issues around religious conflict.

Anke Snoek is a PhD student in the Department of Philosophy at Macquarie University. She has published articles on autonomy, freedom, Agamben, Kafka,

Foucault and addiction and a book entitled *Agamben's Joyful Kafka: Finding Freedom Beyond Subordination* (Bloomsbury 2014).

Ryan Tonkens is a Lecturer at Monash University. His current research interests lie mostly at the intersection of applied ethics and advances in technology, especially biotechnology (for example the ethics of human prenatal genetic alteration), assisted reproductive technologies (for example the ethics of embryo abandonment) and artificial intelligence (for example the ethics of robotic warfare). Ryan is also currently developing a virtue-ethical theory of procreative ethics and the ethics of parenthood. He completed his PhD in Philosophy at York University (Toronto) in 2012, and was a post-doctoral research fellow at Novel Tech Ethics in the Faculty of Medicine at Dalhousie University (in Halifax, Nova Scotia, Canada) before coming to Monash.

Andrés Vaccari is Associate Professor at the National Council of Scientific and Technological Research of Argentina (CONICET). He works at the Centre of Studies on Science, Technology, Culture and Development (Universidad de Río Negro, Argentina), and is Associate Researcher at the Department of Philosophy of Macquarie University (Australia). His areas of research are the philosophy of technology, post-humanism and technological models in biology.

Chapter 1

Introduction

Jai Galliott and Mianna Lotz

The Spartan city-state produced what has been perhaps one of the most ruthless military forces in recorded history, second only to Hitler's Schutzstaffel. Crucial to Sparta's supremacy was the belief that military training and education began at birth. Those judged by state officials to have failed the first round of selection for military service, which began at an inspection in the first few days of life, were left outside the city walls to die of starvation (Lendon 2006, p. 112). In many ways, those who perished were the fortunate ones. To 'harden' the survivors and prepare them for battle, potential Spartan warriors were subjected to extreme temperatures, beatings, sleep deprivation and regular sexual abuse. As with the British, who later did much the same in their military academies to produce the soldiers that would eventually carve out the British Empire, the Spartan regime is renowned for its effectiveness on the battlefield. Those children who completed their military training went on to become some of the most feared warfighters in the entire ancient realm and for much of the time since, politicians and military chiefs longed for technologies that would enable them to avoid the cruelty for which the Spartan regime is now remembered, while still producing effective soldiers who will kill on command, fight without showing signs of fear or fatigue and generally behave more like machine than human beings.

In the absence of means to actualise this desire, it has long been thought that the future of warfare is all about army tanks, fighter jets and missiles. Today, with the advent of unmanned systems that operate across land, sea, air and space, our hopes are attached to the idea that we will soon be able to fight our battles with soldiers pressing buttons in distant command centres. But despite significant investment in what were supposed to be our robotic saviours, much recent warfare has turned out to be a very messy business, leading theorists to question what can be achieved without a human 'in the loop' (Krishnan 2009; Singer 2009; Galliott 2015). Some critics point to the fact that while active combat and reconstruction operations are technically complete, war in the Middle East drags on to this day and is still fought on a human scale in the mud and dust, not with what are typically large and impersonal killing machines (Galliott 2013). While there are most certainly unmanned systems that are useful in the battlefield, enemy forces are now accustomed to fighting in technology-saturated battlespaces and surface only when ready to attack, disappearing into fields and tunnel systems once the skirmish is over. This effectively means that military forces must have 'boots on the ground'. Soldiers are not the cannon fodder of earlier days and must now be highly trained, super strong and have the requisite intelligence and mental capacity to handle the highly complex, dynamic, network-centric military operating environment. It is only as we progress into the twenty-first century that we get closer to realising the Spartan ideal of creating a soldier that can endure more than ever before without – it is hoped – violating human dignity.

Terminator-style weaponry may be many decades or even centuries away, but more realistic efforts to engineer a 'super soldier' are currently under way. We are no longer limited to so-called 'natural' methods of enhancement, whether it be Spartan-style conditioning or simply sending soldiers to the gym. The modern 'military human enhancement' effort draws on the fields of neuroscience, pharmacology, biology, genetics, nanotechnology and robotics. It is fuelled by the United States Army's flagship science and technology initiative, which aims to develop a 'Future Force Warrior' that is highly independent and has superhuman strength (Webster 2012, pp. 98–112). Hundreds of millions of dollars were invested into this program, but it is now largely defunct due to budget measures aimed at ensuring America avoids its ever-looming 'fiscal cliff'. However, it would be short-sighted and perhaps even strategically dangerous to think that military forces have abandoned efforts to upgrade service members' minds and bodies to create the super soldiers that are necessary to match the increasing pace of modern warfare and dominate the growing militaries of the Indo-Pacific region. Slogans such as 'be all that you can be, and a whole lot more' still reign strong in the office of the United States Defense Advanced Research Projects Agency (DARPA) and even in these tough fiscal times, it lodged a FY2015 budget request for more than 300 million dollars of funding for biomedical and biological research under which its 'performance optimization' programs fall (Lin et al. 2014; Department of Defence 2014). DARPA's current unclassified projects focus on: 1) widening physical capabilities by improving strength and mobility with nano-reinforced exoskeletons and other external devices; 2) improving cognitive abilities such as memory, attention and awareness through the use of networked body suits and pharmacological means; 3) enhancing senses such as smell, sight, taste and hearing; and 4) altering the human metabolism to allow for increased endurance, rapid healing and the digestion of otherwise indigestible materials (Lin, Mehlman and Abney 2013; Lin et al. 2014).

It must also be remembered that several emerging powers, including China, Russia, India and the European Union, all have the capacity to acquire and implement the full range of technologies that could lead to the creation of super soldiers (Silberglitt et al. 2006, p. xxiv). The Chinese military human enhancement program is particularly concerning, given that compulsory participation is likely to be mandated. Furthermore, it is reasonable to expect that if China were to develop sophisticated enhancements suitable for wide-scale operational deployment, it would quickly gain military superiority over the United States or any other nation-state, upsetting international order. For this reason, if no other, we should take the opportunity to start asking the difficult normative questions about how – and more to the point whether and on what basis – we should proceed with military human enhancement. The aim of *Super Soldiers: The Ethical, Legal and Social Implications* is to provide the first comprehensive and unifying analysis of the moral, legal and social questions concerning military human enhancement, with a view towards developing guidance and policy that may influence real-world decision-making. Upon close consideration, there is a plethora of questions that demand serious attention. For instance, there are general concerns about the justifications for enhancing soldiers and a range of more specific worries about fairness, the implications for society, the challenges to our traditional conception of medical ethics, risk assessment and design, responsibility, governance and the law. In general, however, it is the tough, practical and forward-looking philosophical questions that are at the core of this volume.

For ease of reference, the chapters of this volume are divided into four parts. Entitled 'What, Why and How', Part I considers how we might define, construct and justify the 'super soldier'. It begins with Chapter 2 from Andrés Vaccari, who looks at the role of agency in an age of synthetic organisms, cyborgs, autonomous robots, human-machine systems and enhanced soldiers. He asks whether we can preserve the language of agency and intentionality. His answer is a hesitant 'yes', as he puts forward a meta-theoretical account that tends towards the post-humanist pole of the many possible perspectives on agency. In Chapter 3, Joseph Pugliese asks us to think about the blurry but important line between human and machine. When is a soldier not a soldier? To shed light on this matter, he proposes an approach that contextualises the relation between bodies and technologies, rather than simply identifying and then grasping onto something that makes us uniquely human. In Chapter 4, Barbara Gurgel and Avery Plaw show that enhancement is not always artificial, external or technological and do so via an exploration of a new frontier in military training that is of great significance in the wake of recent wars in the Middle East: cultural training. They highlight that cultural training is critical if soldiers are to function effectively in foreign environments and avoid incidents of cultural insensitivity, such as burning the Qur'an, humiliating local women and mistreating bodies of the dead. They describe the US military's recent efforts to address these and similar challenges and review several reports from experts in these fields, drawing some preliminary conclusions about what has been and needs to be done. In Chapter 5, Ryan Tonkens takes issue with one of the more common justifications given for the employment of emerging technologies, which is essentially an appeal to military necessity. It holds that military forces and individual soldiers have an obligation to embrace enhancement efforts in order to have a legitimate chance of protecting themselves and the citizens they defend. The main thesis of Tonken's chapter is that the use of soldier enhancements is inconsistent with the long-term goal of peace and that exclusive appeal to military necessity is insufficient to justify such enhancement efforts, even if they are in line with military proficiency.

Part II investigates 'General Problems and Consequences' concerning both present and future military human enhancement endeavours. In Chapter 6, Armin Krishnan explains that it will be difficult for Western democracies to make the transition from a traditional to an enhanced military and that one likely solution for dealing with this challenge is to outsource enhancement functions to private military contractors, who will train and employ small groups of permanently enhanced mercenaries. His chapter discusses the different types of enhancements that could filter through to private military contractors and outlines some of the ethical and legal implications regarding enhancement in this context. In Chapter 7, Robert Simpson discusses technological asymmetry, which poses a problem that has plagued all military technologies at one point or another, from the bow and arrow through to modern-day drones. In brief, when major technological disparities separate the opposing sides in a conflict, there is a plausible case to be made that these gaps render unjustifiable the use of lethal violence by a major military power against a small one. The standard suggestion is that in these sorts of conflicts, major powers ought to eschew warfare in favour of something resembling international policing. However, Simpson argues that a shift to a policing approach cannot be obligatory for the superpower and that any rationale for such argument is critically flawed by the advent of enhancements that alter the risk dynamics of political conflict.

In Part III, we delve into issues of 'Military Medical Ethics'. In Chapter 8, Anke Snoek looks at the relationship between the military, soldiers and synthetic drugs, advancing three key points. The first reveals that the potential for addiction to synthetic drugs is not a mere consequence of the substance itself, and depends on personal characteristics and the context in which the substance is used. The second stipulates that military use of drugs should always be considered from within a military ethics perspective capable of separating decisions to go to war from decisions made in war and individual responsibilities from hierarchical responsibilities. The final point is that the role of the drug user should be taken more seriously in the moral analysis of military human enhancement. In Chapter 9, Steve Matthews examines the biotechnical challenges to moral autonomy and argues that for military human enhancement to be permissible, agents must have the capacity to form psychologically appropriate actions and experiences into a unified morally coherent self conception, all of which is arguably quite important to ethical conduct in war and efforts to rehabilitate soldiers upon their return home. In Chapter 10, Katrina Hutchinson and Wendy Rogers explore the ethical considerations relevant to military surgical innovation. They start by defining surgical innovation, providing a historical survey of surgical innovation in a military context, and then moving on to consider issues such as harm to soldier-patients, informed consent and conflicts of interest.

Finally, Part IV deals with matters of 'Law, Responsibility and Governance'. In Chapter 11, Alex Leveringhaus investigates the attribution of responsibility to enhanced soldiers. He starts by detailing the general nature of responsibility and its links to just war theory, and then looks at the implications of military human enhancement for the moral agency of warfighters and any attempts to impose retrospective and prospective responsibility. In Chapter 12, Seumas Miller explores the implications of emerging technologies for the collective responsibility of humans in war. At the same time, he warns us of the perils of a future in which any system can autonomously conduct lethal operations, suggesting that it would be near impossible for autonomous systems to meet the just war principles of military necessity, discrimination and proportionality. In Chapter 13, Joseph Savirimuthu considers how the International Committee of the Red Cross (ICRC) or International Committee for Robot Arms Control (ICRAC) should respond to the legal and regulatory challenges posed by futuristic warfighters on the rapidly evolving battlefield. He argues that we must properly understand the nature of what he calls the 'problem of disconnection' before we can reflect on the complex interactions between law, technology and policy.

Together, these discussions offer a broad but very thorough analysis of the main ethical and philosophical questions that must be grappled with if we are to move responsibly into an era of enhanced soldiering that will profoundly change the past and current 'landscapes' of war. It is hoped that they will provide the guidance that will enable clear-sighted anticipation of the challenges posed by super soldiers, and thereby help us to avoid developments that violate our most deeply held ethical commitments concerning justice in war and, especially, respect for human safety and dignity.

References

Department of Defence 2014, *Fiscal Year (FY) 2015 Budget Estimates*, Defense Advanced Research Projects Agency, Defense Wide Justification Book, vol. 1, Research, Development, Test & Evaluation, Defense-Wide.

Galliott, J. 2013, 'Unmanned Systems and War's End: Prospects for Lasting Peace', *Dynamiques Internationale*, 8(1): 1–24.

Galliott, J. 2015, *Military Robots: Mapping the Moral Landscape*. Farnham: Ashgate.

Krishnan, A. 2009, *Killer Robots: Legality and Ethicality of Autonomous Weapons*. Farnham: Ashgate.

Lendon, J. 2006, *Soldiers & Ghosts: A History of Battle in Classical Antiquity*. New Haven, CT: Yale University Press.

Lin, P., Mehlman, M. and Abney, K. 2013, *Enhanced Warfighters: Risk, Ethics, and Policy*. San Francisco, CA: Greenwall Foundation.

Lin, P., Mehlman, M., Abney, K. and Galliott, J. 2014, 'Super Soldiers (Part 1): What is Military Human Enhancement', in S. Thompson (ed.), *Global Issues and Ethical Considerations in Human Enhancement Technologies*. Hershey: IGI Global.

Silberglitt, R., Anton, P., Howell, D. and Wong, A. 2006, *The Global Technology Revolution 2020, In-Depth Analyses*. Arlington: RAND National Security Research Division.

Singer, P. 2009, *Wired for War: The Robotics Revolution and Conflict in the 21st Century*. New York: Penguin.

Webster, J. 2012, *What Brings a Soldier to His Knees*. Edinburgh: Westbow Press.

PART I
What, Why and How

Chapter 2

Abjecting Humanity: Dehumanising and Post-humanising the Military

Andrés Vaccari

Ethical questions concerning war are often posed around notions of the human – humanness, humaneness, humanity – and its shadows: the inhuman, dehumanised, post-human. There are two questions here, one metaphysical and one ethical. The first one is about the ontological limits of the human in a novel military context populated with new actors who straddle the categories of human and machine, biology and technology. As way of an answer, I offer a framework to consider a broad range of (real and prospective) hybrids with military applications, such as synthetic organisms, cyborgs, autonomous robots, human-machine systems and modified humans in general. My aim is to link two problems: human agency and the ontology of new hybrids. The metaphysics of agency (particularly, in relation to intentionality) has been the traditional basis on which to distinguish natural from made, human from nonhuman – and this is the central link between the two questions.

The ethical question is closely tied to this, and concerns moral agency and responsibility. Much of the philosophical literature on new military technologies is concerned with this, and particularly with the legal approaches that should be developed to deal with these novel realities. This ethical aspect comes to rest on metaphysical matters. To ask about responsibility is to pose the metaphysical problem of the author of an action – the question of agency. And agency has become increasingly hard to trace in an age in which humans predominantly carry out their activities in the midst of 'ever more distributed and entangled socio-technical systems' (Simon 2014, p. 2). According to the Office of the US Air Force Chief Scientist, this trend will accelerate in the next decades:

> … natural human capacities are becoming increasingly mismatched to the enormous data volumes, processing capabilities, and decision speeds that technologies either offer or demand. Although humans today remain more capable than machines for many tasks, by 2030 machine capabilities will have increased to the point that humans will have become the weakest component in a wide array of systems and processes. Humans and machines will need to become far more closely coupled, through improved human-machine interfaces and by direct augmentation of human performance (Technology Horizons, 2010: ix–x).

The loss of human agency and the increasing autonomy of systems are two aspects of the same phenomenon. Can we preserve the language of responsibility, intentionality and a unique human autonomy while situating agency in this dense realm of technology, sociality and materiality? And how does the nature of warfare frame and shape these issues?

The framework I propose is meta-theoretical. It draws a continuum between two poles, mapping possible perspectives on agency: *humanist* and *post-humanist*. The humanist pole represents liberalist or 'modernist' conceptions of humanness and agency; while the post-humanist end of the spectrum captures systemic perspectives that see agency as the product of couplings of human and nonhuman components. My framework is based on the post-humanist perspective. I will use the term 'hybrid' to encompass all possible entities, from the fully synthetic to the biologically modified. It does not matter whether these hybrids are present or future, real or imagined, possible or unfeasible.

Although my sympathies tend toward the post-humanist end of the spectrum, I will not argue for or against any perspective. The last section of this chapter consists of a dialogue between two characters who discuss the ethical implications of their respective theoretical positions. The dialogue also brings the discussion into the specific, current military context. The aim is broad and exploratory, and this chapter attempts to map out the ontological and ethical choices we have in this debate, in terms of thinking agency in complex systems.

Framing Agency: Humanist and Post-humanist Approaches

Our first task is to unpack these two contrasting theories of agency, and how they board the human, technology and their relations. The two poles do not exhaust all possible takes on the issue, but characterise two broad analytical frames that may complement each other in some versions, and on which there are considerable disagreements and variations. I will begin mapping out this spectrum at the humanist end, with liberalist-rationalist ideas about agency, humans and technology.

Humanism

According to the classical definition of Anthony Giddens (1984), action 'depends upon the capability of the individual to 'make a difference' to a pre-existing state of affairs or course of events' (p. 14). An agent is characterised by a causal power – and not just any power but one that is exerted in view of intentional aims. In this manner, the question of agency is closely intertwined with some of the trickiest problems of philosophy: intentionality, freedom, consciousness, control, representation, autonomy and power. In a strong humanist reading, agency requires the capacity to act self-reflexively: a full agent is someone who is able to *represent* his/her own states to him/herself, along with the relevant features of the 'worldy' context, and is able to act upon the world in order to change a state of affairs and bring it in line with goals set beforehand.

A strong understanding of agency requires a concomitantly strong, Kantian notion of autonomy in which the agent chooses his/her goals freely. Agency is the capacity to self-determine our own ends and therefore our own selves; this is the link between freedom and autonomy.

One problem for such a strong understanding is that few of our actions would satisfy what it demands. This is particularly visible in organisational milieus, where goals are already set as internal functions of systems and dictated by their structural necessities. Alternatively, we could argue that the difference between autonomy and non-autonomy does not coincide with that between agency and non-agency (Buss 2013). Heroin addicts and people who act compulsively (according to reasons they do not assent to) are typical examples of non-autonomous agency. They are actors who do not freely perform their actions according to reasons they agree with, yet they are *agents* of their actions; by which we mean they are the main cause of bringing about a certain state of affairs.

In all cases, intentionality establishes the minimal threshold for agency. This leaves out automatic or non-intentional acts, such as those performed by machines, invertebrates and by humans (arguably most of the time). An automatic machine may be autonomous in terms of its 'capacity to operate in the real-world environment without any form of external control, once [it] … is activated and at least in some areas of operation' (Lin, Bekey and Abney 2008). Indeed, it is acceptable to speak of these types of machines as quasi-agents in this sense. But *real* intentional agents, according to this view, require autonomy in the Kantian sense: a machine 'could not become a *morally* autonomous "law unto itself" and serve its own ends; hence it cannot be morally responsible for its actions' (p. 65). Agency, in this view, is always already moral.

The structure of intentional agency is manifested in *action*: the concrete, real-time unfolding or expression of agency. Intentionality is the main criterion to distinguish actions from *events*, or agent-causality from mere physical causality.[1] According to the classical picture of means-ends rationality, an action only takes place when an end has been intentionally established previously, as the result of deliberation. This also serves to distinguish actions from other forms of behaviour, such as those Mark Rowlands (2006) calls 'deeds': acts requiring no prior intentions or representational content. Once the goal is set, the agent deliberates on the appropriate means to reach that end state. This is where technology comes in: as *means*, the instrumental aspect of action.

Action has two distinctive features: it is irreducibly *teleological* and *normative* in structure and in nature. Even though an action might not accomplish what it intended, the behaviour is goal-driven and sets its own (normative) conditions of success.

---

1   There are two problems that are not our concern here: mental causation and the issue of defining *basic* actions.

In turn, these views on agency and action come to rely on a philosophical anthropology. As Stephen Fuchs writes: 'Agency essentialism thinks of agency, intentionality, and mind as something persons have *qua* persons' (2001, p. 32). In other words, humanists understand agency as a distinctively human property that relies on mental capacities such as introspection, intentionality, willing, self-reflexivity and future projection. Even though humans were deprived of all capacity for action (say, by being completely paralysed), we can say that agency remains as a power or potentiality; or, alternatively, we could hold the view that an agency-less entity lacks humanness or personhood.

An agent, then, is an individual substance with distinctive causal powers that distinguish it from other natural agents (for example, chemical substances) (Lowe 2010). This entails that an agent should be considered a *unified* thing, as Christine Korsgaard argues: 'it is essential to the concept of agency that an agent be unified. ... For a movement to be my action, for it to be expressive of *myself* in the way that an action must be, it must result from my entire nature working as an integrated whole' (2009, pp. 18–19).

These perspectives involve not just a philosophy of the human but also a philosophy of technology. Two central theses flow from the above: technology is *instrumental* and *intentionalist* in character. In action theory, this is evident in the numerous thought experiments involving artifacts of some kind. Here is a case from Donald Davidson:

> A man may try to kill someone by shooting at him. Suppose the killer misses his victim by a mile, but the shot stampedes a herd of wild pigs that trample the intended victim to death. Do we want to say the man killed his victim *intentionally*? The point of the example is that not just any causal connection between rationalizing attitudes and a wanted effect suffices to guarantee that producing the wanted effect was intentional. The causal chain must follow the right sort of route (2001, p. 78).

The main relevance of technical means in these accounts is that they introduce external causal chains potentially outside of the subject's agential control. To say that an action was performed intentionally means it was carried out appropriately, according to normative conditions set by the act of intending. It follows that technologically mediated actions ('technical' action and agency) do not require a specific analytical framework. It is almost irrelevant whether an action is carried out with or without tools, or within a dense system of causal mediations; technology may complicate but does not modify the basic structure of intentional action. Interestingly, bodily movements present an analogous problem to instruments in some theories of agency, since they lead to the same 'chasing' of causes (Davidson 1963).[2] The main issue is causation: what does it mean to say that an agent is the

---

2   The issue here (which, of course, we will not go into) is between models of agency that see mental states or events as the prime cause of action, and agent-causal theories

cause of an action? The question of responsibility becomes a matter of tracing the tortuous paths of causes and effects, from intentions (or mental events, or unified agents) to results.

The instrumentalist sees artifacts as metaphysically clean channels for the will of human beings; his motto is 'It is not the technology but who uses it'. In turn, this implies the *value-neutrality* thesis. We can see how intentionalism, the second predominant position in humanist philosophies of technology, is linked to (and to an extent presupposed in) instrumentalism and value-neutrality. The humanist does not offer an ontology of hybrids, but an ontology of the natural-artificial in which these two realms are distinguished according to clear criteria. The distinction between natural and artificial objects hinges on the fact that the latter have been produced with a purpose or function, while the former are the result of blind natural forces. Thus, artifacts are ontologically defined as matter organised in accordance with intended function, and issues around the ontology of artifacts should be subsumed under the more general problem of intentionality (Dipert 1995; Hilpinen 2004; Baker 2004; Thomasson 2007). Both instrumentalism and intentionalism rely to various degrees on a distinction between intentionality and materiality, leading to what Beth Preston calls the *centralised control model* of action. This model has two main features: an emphasis on individual action and planning, and a reliance on a model of production in which forms are 'impressed' on matter (2013, pp. 15–43). Again, agency and ontology are intertwined: it is intentionality and agency, manifested in use and production, that serve to distinguish the artifactual as a derivative ontological class.

Even vast-scale technological systems show this derived or second-order intentionality. Intentions are realised as *designs* that expand the capacities and existential opportunities of a particular subject, be it individual or collective; in this manner, the world of artifacts 'produces the enlargement and opening up of the space of accessible opportunities' (Broncano 2009, p. 67), and this is an expression of a characteristically human 'dimension of freedom: that which is linked to the imagination of desirable alternatives' (p. 67).

Post-humanism

At the other end of the spectrum, we find *post-humanist, systemic* or *externalist* accounts of agency. These proposals have come from diverse disciplines and research programs, ranging from science and technology studies to the philosophy of mind, and from archaeology and sociology to evolutionary theory. Despite its heterogeneity, this work clusters around some common themes. Approaches such as actor-network, post-phenomenology, extended cognition, material agency, assemblage theory, cyborg anthropology and material culture studies, to cite some of the most prominent, share a key concern with framing the relation between

---

that argue that the whole agent is the cause. See Davidson (2001, pp. 3–20) for a classic introduction to some of these problems.

intentionality and world in a way that avoids the trappings of Cartesianism and the agency-structure debate. Among other things, it has been suggested that properties formerly exclusive to humans, such as mind and agency, should be afforded also to the external constituents of action; this could encompass artifacts and technologies, embodied aspects of cognition, the physical and informational features of the surrounding environment and even texts and bodies of knowledge. According to this view, human and nonhuman entities are found inextricably composed into ontologically hybrid networks or 'assemblages' (Deleuze and Guattari 1987) in which nonhuman actors exert a veritable agential role that can radically transform, and even give rise to, intentions and actions.

In cognitive science and the philosophy of mind, this translates into the idea that internal cognitive capacities and external supports are coupled dynamically, forming a distributed cognitive system that stretches beyond the traditional perimeter of the mind, as bounded by the skull (for example, Hutchins 1995; Clark and Chalmers 1998; Clark 2008; Menary 2010). In Andy Clark's terms, we have always been 'natural-born cyborgs' (2003) with a biological and cognitive plasticity 'naturally' permeable to coupling with artifacts and the environment. Following Katherine Hayles, I will adopt the term 'post-humanism' as a way to come to terms with this historical and cultural condition marked by the breakdown of any 'essential differences or absolute demarcations between bodily existence and computer simulation, cybernetic mechanism and biological organism, robot teleology and human goals' (1999, p. 3).

Post-humanists tend to reject human-centred analyses that consider the human as a fixed individual substance with distinguishing properties. Jean-Marie Schaeffer has elegantly called this shift 'the end of the human exception' (2009), a loss of ontological uniqueness that, he argues, is a logical outcome of naturalistic interpretations of human nature.

Going back to Giddens' definition, we can see that the systemic thesis fits well with one central condition for agency: bringing about a change in a state of affairs. Material entities 'have causal agency' in this sense since they 'co-constitute real-time activities of human beings' (Kirchhoff 2009, p. 212). In other terms, actors such as cognitive ecologies and the physical features of artifacts play ineliminable causal roles in the outcomes of action (Kirchhoff 2010). Agency is, then, an emergent and distributed feature of a system assembled from ontologically heterogeneous elements; it is not the property of a pre-existing subject distinguishable from the material and bodily conditions in which she is embedded.

One consequence of sharing agency across the participants of action is that agents become a function of action and exhaust themselves upon its completion; there is no 'agency' as a substantial quality of an entity that governs, structures and supervises action. The very notion of action is weakened and diffused, becoming a shifting real-time arrangement – something closer to an event. However, action still plays an important role in understanding agency; its teleological and normative features are reformulated in biological terms, as the capacities of *living organisms*, manifest in the behaviours of the simplest lifeforms (a bacterium swimming up a sugar gradient is the most popular example in the literature). This biological reformulation of autonomy has the consequence of bringing the living and the technical close together; autonomous machines do not merely imitate life but come to supplant it, in a way.

Problems around agency naturally drift to ontological considerations; in this case, agency is conceived a transversal articulation of nonhumans and (sometimes) humans. The metaphysical specificities of biological and technological systems are blurred in two main senses: (1) in terms of their physical and ontological boundaries; and (2) in terms of their defining properties.

The philosophy of technology is our last stop in this survey. In post-humanist accounts, technology is removed from the realm of means (mere instruments) to become a significant, constitutive shaper of human existence. Peter-Paul Verbeek (2005) draws a distinction between two dimensions: a *hermeneutical* dimension in which 'artifacts mediate human experience by transforming perceptions and interpretive frameworks, helping to shape the way in which human beings encounter reality'; and an *existential* dimension in which artifacts give 'concrete shape to their behavior and the social contexts of their existence' (2005, p. 195).

The notion of mediation acts as a common framework to think about technology in both cognitive and agential dimensions. Bruno Latour (1999) illustrates this notion of mediation with the example of a gun and a 'citizen'. Firstly, the combination of these two actors produces two dominant stories about technology:

The first story is: *Guns kill people*. This is a materialistic, substantivist story: 'the gun acts by virtue of *material* components irreducible to the social qualities of the gunman' (Latour 1999, p. 176). The moment he holds the gun, a good person might become dangerous, part of a 'script' (Akrich 1992), a 'dynamic space' (Deleuze and Guattari 1987, p. 404) or program of action built into the artifact.

The opposite story is: *Guns don't kill people*; people *kill people*. This is a sociological, humanist account in which the weapon is 'a tool, a medium, a neutral carrier of the human will' (Latour 1999, p. 177). The gun simply accomplishes more efficiently a goal that already existed as the content of human intention.

These two accounts may be respectively termed substantivist and instrumental, and the concept of mediation is meant to cut a middle path between them. According to Latour, the mistake of both humanist and substantivist accounts is 'to start with essences, those of subjects *or* those of objects' (1999, p. 180). Latour argues that this transformation is symmetrical: you (literally) become another person when you hold the gun *and* the gun becomes another thing in your hand. This reciprocal change is called *translation*: the creation of a new goal out of the meeting of two actors: a 'displacement, drift, invention ... the creation of a link that did not exist before and that to some degree modifies two elements or agents' (1994, p. 32).

Peter-Paul Verbeek argues that these considerations change the moral status of technologies. Artifacts are moral agents inasmuch as they open up possible spaces of intentionality and 'actively co-shape people's being in the world' (2006, p. 364). Intentionality is 'the directedness of human beings toward their world' (2008, p. 13); yet there is no 'pure' and unmediated action or experience that is not already technologically constituted. Taking the obstetric ultrasound as a case study, Verbeek examines how this technology shapes the perceptual presence of the object (the unborn child) and the moral space of decision-making. Particular attention is given to the material qualities of representation: the fact that, in ultrasound images, the foetus appears in a much larger size and as an object that is independent of the mother's body. Ultrasound imagery *ontologically constitutes* the foetus (that is, it is not merely an *interpretation*) as an individual person and as medical subject (*patient*) (pp. 15–16). It also shapes the mother, father and unborn child in specific ways, and in terms of their relations (p. 17). *Having* a child has been translated (in the Latourian sense) as *choosing* to have a child. The post-phenomenological account concludes that 'ethics is not solely a human affair, but a matter of associations between humans and technologies' and thus cannot depart from a separation between them (p. 18).

Ontological and Ethical Issues Pertaining to New Military Actors: A Concluding Dialogue

Now, where does all this leave our super soldiers, warrior hybrids and ethical paradigms? To explore these issues, I will hand over the discussion to two characters representing humanist and post-humanist viewpoints. They are, respectively, Plato, who needs no introduction; and Roy Batty, the android who spearheaded the 'replicant' insurrection in Philip K. Dick's novel *Do Androids Dream of Electric Sheep?* (1968), and who Rutger Hauer unforgettably portrayed in *Blade Runner* (1982, dir. Ridley Scott), the film based on the novel. My aim is not to settle the score either way but to explore the ways in which these two approaches deal with the ontological and ethical questions of new military actors, and of the military context in general.

Roy Batty: The heuristic framework I propose allows us to capture a broad range of real and imagined natural-artificial hybrids. For example, a World War I soldier with a rifle and a super soldier produced by synthetic biology can be regarded from the same perspective: as a system that combines functions that are ontologically indistinct in terms of their natural or artificial origin – concepts that are materially meaningless. It follows that the distinctions exogenous-endogenous, addition-modification and essential-accidental do not apply, since any intervention becomes a structural-functional feature of a whole new assemblage. In this scheme, modifications to existing biological systems and the engineering of autonomous weapons from scratch are essentially the same thing: the creation of an artifact-organism, a body that should be classified not by its nature but by its specific *powers*. We post-humanists are not interested in what a thing *is* but in what it *does*. AWS are lifelike inasmuch as their patterns of action show dimensions of teleology and normativity, a form of biological autonomy common to both organisms and sufficiently complex machines. It also follows that modifications do not add, expand, enhance or improve on something that was already there and which can be used as a normative benchmark (mind, body, person, machine, human).

Plato: Let me interrupt you there. This supposedly causal contribution of nonhumans to the powers of an assemblage complicates assigning agency to any specific actor. And this has some horrific ethical consequences. Your views seem to muddle up the questions of moral agency and responsibility.

Batty: Or at least to pose them in different terms. We humans are responsible, yes, but not *just* for our actions.

Plato: So human beings have no essence, which means that any human modification would change the ontological and moral status of the person, perhaps reducing it to a bundle of cognitive and physiological capacities. In contrast, we humanists have a clear normative yardstick to distinguish humans from things, and natural from artificial objects. We concede that it is hard to trace chains of intentions and actions across large organisations, let alone in the fog of war. But the ethical philosopher is concerned with boundaries chiefly because laws are based on the notion that we can distinguish moral agents from their instruments. Furthermore, humanness is the basis of the 'principle of humanity' as entrenched in international law. How are we to conceive of 'human rights' without some notion of humanness? Even if you were right, we could still *choose* to reconstruct the causal history of an action in this framework, since the alternative you propose is horrifying. And of course I do agree that boundaries are not neutral and objective. To quote from one of your favourite post-humanists, Karen Barad: 'boundaries are interested instances of power, specific constructions, with real material consequences' (1996, p. 182). This is precisely why we need to impose these boundaries, because of the associated human consequences.

Batty: To us, the ethical question is about concrete action in the face of specific agential assemblages. I'll quote Barad back to you: 'The acknowledgment of 'nonhuman agency' does not lessen human accountability; on the contrary, it means that accountability requires that much more attentiveness to existing power asymmetries' (2007, p. 218f). To us, the artifacts themselves are morally questionable, so the focus of intervention should be the *design*. We need to enable 'designers to actively anticipate the morally relevant role of technology' (Verbeek 2008, p. 25). We need to think up new *philosophies* of design.

Plato: I, on the other hand, believe there is no machine responsibility, only product liability. Design malfunctions can ultimately be traced back to human failure. So, the answer is to regulate the *use* of technologies, not their design. The morality of a machine is, at best, an *operational morality* (Lin, Bekey and Abney 2008, p. 26). Ronald Arkin (2010), in fact, has famously argued that we should give more ethical autonomy to unmanned systems, since 'they can perform more ethically than human soldiers' (p. 334). Existing legal categories should be expanded and accommodated to deal with new cases. For example, bio-enhanced soldiers may be considered 'biological weapons' under the Biological and Toxin Weapons Convention (Lin, Mehlman and Abney 2013, pp. 8–9). Lin, Bekey and Abney also discuss the law of agency, which considers cases 'in which the power of agency is transferred between parties' (2008, p. 59). In these instances, legal agency is distributed, since the agents enact legal powers from afar. They suggest that this could be applied to autonomous machines. As you can see, we humanists have plenty of conceptual ammunition to come to terms with these hybrids.

Batty: The politics and ethics of technology are distributed in more complex and subtle ways than what humanists suggest. As Lucas Introna says, we must get a grip on 'who (in terms of human and non-human actors) is doing what, when and how, i.e. we need to get a grip on the problem of the on-going constitution (or constitutive conditions) of sociomaterial agency' (2014, p. 33). In this regard, Judith Simon (2014) introduces a useful distinction between *responsibility* (which requires intentionality) and *accountability*, which applies to artifacts and systems. The problem is organisational: how we design systems so that people become *responsibilised*. It is the assemblage what creates the *identities* of the human agents and positions them as moral agents in a particular relation to their actions and practices (Introna 2014). Barad puts it nicely: 'We (but not only "we humans") are always already responsible to the others with whom or which we are entangled, not through conscious intent but through the various ontological entanglements that materiality entails' (2007, p. 393). Consider the clichéd case of the mad dictator who gets hold of an army of AWS. Whereas you would hold the dictator responsible, I would ask: where did he get the machines from? What network enabled him to perform the actions? Shouldn't the whole international weapons trade system be equally condemned? Let me give you another example. The intended function of AWS is a reduction in the number of casualties on the deployment side and perhaps also of non-combatants on enemy grounds. Yet the widespread introduction of AWS could also lower the political costs of military conflict, thus making war 'a preferred or convenient method of conflict resolution' (Lin, 2010, p. 313; Sparrow 2009). In what regards remotely operated drones, cultural perceptions that they are an ignoble manner of conducting war may have swelled the ranks of insurgents in Afghanistan and elsewhere.[3] It has also been said that this technology may cause remote operators to be trigger-happy. These are not unintended consequences, but an aspect of the agency of technologies. We often pose the problem in human-centred terms: technology is out of our control, or it controls us. But systems are always in control, in the strict cybernetic sense; they self-organise, actively shape their own boundaries and regulate their own activity.

Plato: But your perspective is too broad to provide any useful guidance.

Batty: I disagree. See, you and I are already part of the network of war and ethically implicated in it. If you allow me, the following piece, from the satirical newspaper *The Onion* will make a good case study:

---

3   Andrew Kilcullen, former adviser to David Petraeus (US Army General, now retired, and ex-Director of the CIA) on counterinsurgency, said each innocent victim of a drone strike 'represents an alienated family, a new revenge feud, and more recruits for a militant movement that has grown exponentially as drone strikes have increased' (cited in Hasan 2010).

October 9, 2013: Military Unveils Bionic Super-Soldiers Capable of Withstanding Mental Toll of War

Touting them as the next stage in modernized combat, representatives for the United States military unveiled today a new line of bionically enhanced 'super-soldiers', capable of withstanding the enormous mental toll of war. Commanders introduced the next-generation, biologically modified troops at a press conference in Washington, telling reporters that the elite military personnel have been engineered to mentally withstand limitless amounts of violence and bloodshed on the battlefield, which would then prevent them from experiencing future bouts of paranoia, anxiety, and crippling depression. …

According to members of the armed forces, the bionic infantrymen are surgically outfitted with an impenetrable mental barrier through which terror, sorrow, guilt, and despair cannot pass. They are reportedly capable of all the functions of a standard infantryman, but with an augmented resilience to psychological scarring resulting from the types of pain and anguish no normal human being should ever experience for prolonged periods of time.

Furthermore, sources confirmed these advanced new conscripts are imbued with the unique ability to withstand the butchery of innocent civilians and the abrupt deaths of their squadmates without these episodes haunting them every day for the rest of their lives.

Despite its humorous guise, the piece has a serious sting. It states two truths: war is an essentially dehumanising enterprise and, in a military context, an 'enhancement' could mean the opposite of what it normally stands for. The joke turns bitter when we consider that in 2012 more US army personnel committed suicide than perished in the war in Afghanistan (Pow 2012). In the case of Australian troops, the number of veterans taking their own lives has tripled the Afghanistan combat toll (Brown 2014). Does it make any sense to speak of a more humane pursuit of war? Or does war, by its very nature, require dehumanisation: the impoverishment of certain capacities essential to humanness, and the reduction of people (soldiers and enemy) to something less than human – an object, instrument, machine? In training to wage battle, soldiers are put through 'inhuman' conditions in order to build up physical and emotional endurance. And in war it is not only acceptable but indeed necessary to treat military personnel and the enemy in ways that humanist sensibilities find morally abhorrent. Dehumanising the enemy is a standard way of establishing a psychological distance that morally excludes the other from pity or consideration (Haslam 2006). Nick Haslam, in fact, has identified *mechanistic dehumanisation* as a specific form of distancing. In this regard, it is not a coincidence that the advent of the modern army is contemporaneous with the rise of mechanicism in the sciences, and the instauration of mechanical forms of production, in sixteenth-century Western Europe.

Plato: Yes, warfare presents us with a distorted moral universe. But, if I read you correctly, you seem to be suggesting that we naturalise dehumanisation and regard it as ethically unproblematic, since it has always been a feature of war.

Batty: Dehumanisation and warfare are very ethically problematic, but on a deeper level. *Humanness* is not a good basis for a whole ethical approach, since the human is essentially a constitution of discourses, technologies and practices. This does not mean that there is no reality 'out there'; only that modal weights and powers of substances are always found commingled in systems. The status of humanness can be taken away from you, or any of us, at the flick of a finger, since it is a function of discourse. Discourses about the human have served, at various stages in history, not to universalise but to exclude, to draw boundaries between a group and a dehumanised other; women, savages and slaves, most notoriously; but also right now, in Guantanamo and in anonymous torture chambers across the Middle East.

Plato: I find it curious that, despite your talk, you still cling to recognisably humanist narratives. Humans have the power to reflect and effect changes in systems in view of certain goals and values. Post-humanism gives us old truisms in new packaging: the world is partly constitutive of subjectivity, agents must often make decisions in circumstances that exceed their competence, technical means often encumber the assigning of moral blame and so on. As Mark Peter Jones (1996) argues, insisting on the agency of things is to commit a category mistake (p. 305). Also, the military has funded some research on sociotechnical systems theory (for example, Walker et al. 2007). This approach acknowledges that deterministic behaviour is difficult to impose on systems of interlinked entities that do no maintain a fixed state (*open* as opposed to *closed* systems – such as those encountered in asymmetrical warfare). We can approach complexity without giving in to this post-humanist mumbo-jumbo. Standards of efficiency and rationality can be maintained by seeking new paradigms of command and control.

Batty: I would argue that, by speaking of laws, norms and ethical frameworks *you* are the one who has already naturalised the inhuman. You can only assign responsibility after the fact, leaving moral agency as a pristine, untouched 'natural' order, as though it was some kind of ecological reserve. Ironically, your approach leaves most of the actors of warfare outside the loop of responsibility (for example, weapons manufacturers and ethicists).

Plato: Perhaps Heidegger was right, and only a god can save us.

Batty: What more suitable way to finish off this veritable Socratic combat than with a myth?

Plato: If this was a Socratic dialogue, I'd be winning.

Batty: Deleuze and Guattari start off their 'Treatise on Nomadology' (1987, p. 351) with the following axiom: *The war machine is exterior to the state apparatus.* Drawing from philologist Georges Dumézil's comparative studies of Indo-European myth, they postulate that political sovereignty consists of two poles or heads, the *magician-king-despot* and the *jurist-priest-legislator*. These two 'antithetical and complementary' functions organise the workings of the State apparatus: the despot binds and rules; the legislator organises and regulates (Dumézil cited in Deleuze and Guattari, p. 351). But there is a third function that the State must appropriate: the warrior, military or war machine. Deleuze and Guattari argue that the war function remains *alien* to the State, the activity of nomads and tribes. The war machine brings its own patterns of thought, codes and ways of organising and occupying space. This means there is an essential tension between political and military power. The war machine 'seems to be irreducible to the State apparatus, to be outside its sovereignty, and prior to its law: it comes from elsewhere. ... [It] is of another species, another nature, another origin ... ' (p. 352).

In their efforts to ethically tame the warrior and the tremendous violence of the war machine, liberal democracies have coded the rules of warfare, mainly in the form of international conventions that define the limits of just engagement and use of force. Among other things, these conventions aim to keep war 'humane', protect the lives of innocent non-combatants and hold officers and soldiers morally accountable for their actions. In addition, the military has established its own parallel judicial structure composed of tribunals, procedures and internal codes of conduct that reflect military values, such as honour and duty. However, it is as though liberal democracy has had to tolerate the war machine in spite of itself, as a necessary evil that must be accepted only for the sake of a greater good – one of the reasons why many people consider 'military ethics' a contradiction in terms. The war machine is a limb of political power, acting in the pursuit of the goals of the liberal State: the protection of its citizens, borders, liberties and economic interests. Yet, the primordial and essential violence of the war machine is clearly discernible under the humanist mask that liberalism has tried to place on it.

Julia Kristeva's theory of abjection (1982) offers another, complementary angle to think about this relationship. What if we think of the state as a subject, a body? In this case, the abject of war stands as exterior to the subject but constitutive of its borders; it is not an object but a 'jettisoned object, ... radically excluded' that draws the subject 'toward the place where meaning collapses' (p. 2). The abject is 'a sickness at one's own body' (Grosz 1989, p. 78) that is forever outside the symbolic order and the reach of reason, yet it is not wholly unconscious: it surfaces as an intrinsically corporeal sign: 'repugnance, disgust, abjection' (Kristeva 1983, p. 11). What is war, then, but the abjecting of humanity, an intolerable yet necessary aspect of the constitution of the liberalist state?

Plato: All very poetic but, if you'll excuse me, I have some more urgent problems to attend.

Batty: Nice speaking to you again.

Plato: I think I'm going to have nightmares tonight.

References

Akrich, M. 1992, 'The de-scription of technical objects', in W.E. Bijker and J. Law (eds), *Shaping Technology/Building Society*. Cambridge, MA: MIT Press.

Arkin, R.C. 2010, 'The case for ethical autonomy in unmanned systems', *Journal of Military Ethics*, 9(4): 332–41.

Baker, L.R. 2004, 'The ontology of artifacts', *Philosophical Explorations*, 7: 99–111.

Barad, K. 1996, 'Meeting the universe halfway: Realism and social constructivism without contradiction', in L. Hankinson Nelson and J. Nelson (eds), *Feminism, Science, and the Philosophy of Science*. Dordrecht: Kluwer Press.

Barad, K. 2007, *Meeting the Universe Halfway: Quantum Physics and the Entanglement of Matter and Meaning*, Durham, NC: Duke University Press.

Blade Runner 1982, motion picture, Warner Bros, USA. Directed by Ridley Scott.

Broncano, F. 2009, 'El mito de la máquina y la agencia técnica', in D. Lawler and J. Vega (eds), *La respuesta a la pregunta: metafísica, técnica y valores*. Buenos Aires: Biblos.

Brown, R. 2014, 'Soldier suicide: Number of veterans taking own lives more than triples Afghanistan combat toll', *ABC News Online*, viewed 11 June 2014, http://www.abc.net.au/news/2014-04-22/number-of-soldiers-committing-suicide-triples-afghan-combat-toll/5403122.

Buss, S. 2013, 'Personal autonomy', *Stanford Encyclopedia of Philosophy*, viewed 13 January 2014, http://plato.stanford.edu/entries/personal-autonomy.

Clark, A. 2008, *Supersizing the Mind: Embodiment, Action, and Cognitive Extension*, Oxford and NY: Oxford University Press.

Clark, A. 2003, *Natural-born Cyborgs: Minds, Technologies, and the Future of Human Intelligence*. Oxford and NY: Oxford University Press.

Clark, A. and Chalmers, D.J. 1998, 'The extended mind', *Analysis*, 58(1): 7–19.

Davidson, D. 1963, 'Actions, reasons, and causes', *Journal of Philosophy*, 60(23): 685–700.

Davidson, D. 2001, *Essays on Actions and Events*, 2nd ed.. Oxford: Oxford University Press.

Dick, P.K. 1968, *Do Androids Dream of Electric Sheep?*, NY: Doubleday and Company.

Deleuze, G. and Guattari, F. 1987, *A Thousand Plateaus: Capitalism and Schizophrenia*. Minneapolis, MN: University of Minnesota Press.

Dipert, R. 1995. 'Some issues in the theory of artifacts: Defining 'artifact' and related notions', *The Monist*, 78(2): 119–36.

Fuchs, S. 2001, 'Beyond agency', *Sociological Theory*, 19(1): 24–40.

Giddens, A. 1984, *The Constitution of Society: Outline of the Theory of Structuration*, Cambridge, UK: Polity Press.

Grosz, E. 1989, *Sexual Subversions: Three French Feminists*, Boston, MA: Allen & Unwin.

Hasan, M. 2010, 'US drone attacks are no laughing matter, Mr Obama', *The Guardian*, 28 December, viewed 21 April 2014, http://www.theguardian.com/commentisfree/cifamerica/2010/dec/28/us-drone-attacks-no-laughing-matter.

Haslam, N. 2006, 'Dehumanization: An integrative review', *Personality and Social Psychology Review*, 10(3): 252–64.

Hayles, N.K. 1999, *How We Became Posthuman: Virtual Bodies in Cybernetics, Literature, and Informatics*, Chicago, IL: University of Chicago Press.

Hilpinen, R. 2004, *Artifact*, Stanford Encyclopedia of Philosophy, viewed July 23 2009, http://plato.stanford.edu/entries/artifact.

Hutchins, E. 1995, *Cognition in the Wild*. Cambridge, MA: MIT Press.

Introna, L.D. 2014, 'Towards a post-human intra-actional account of sociomaterial Agency (and morality)', in P. Kroes and P-P. Verbeek (eds), *The Moral Status of Technical Artefacts*. Dordrecht: Springer.

Jones, M.P. 1996, 'Posthuman agency: Between theoretical traditions', *Sociological Theory*, 14(3): 290–309.

Kirchhoff, M.D. 2009, 'Material agency: A framework for ascribing agency to human culture', *Techne*, 13(3).

Kirchhoff, M.D. 2010, 'Pressing agency beyond the flesh: Three programmatic arguments grounding the notion of material agency', *CEPHAD 2010 // The Borderland Between Philosophy and Design Research*, 1: 91–6.

Korsgaard, C.M. 2009, *Self-constitution: Agency, Identity and Integrity*. New York: Oxford University Press.

Kristeva, J. 1982, *Powers of Horror: An Essay on Abjection*. New York: Columbia University Press.

Latour, B. 1994, 'On technical mediation', *Common Knowledge*, 3(2): 29–64.

Latour, B. 1999, *Pandora's Hope: Essays on the Reality of Science Studies*, Cambridge, MA, Harvard University Press.

Lin, P. 2010, 'Ethical Blowback from Emerging Technologies', *Journal of Military Ethics*, 9(4), 313–31.

Lin, P., Bekey, G. and Abney K. 2008, *Autonomous Military Robotics: Risk, Ethics, and Design*, San Luis Obispo, CA: California Polytechnic State University.

Lin, P., Mehlman, M.J. and Abney K. 2013, *Enhanced Warfighters: Risk, Ethics and Policy*, San Luis Obispo, CA: California Polytechnic State University.

Lowe, E.J. 2010, 'Action Theory and Ontology', in T. O'Connor and C. Sandis (eds), *A Companion to the Philosophy of Action*. Oxford: Wiley-Blackwell.

Malafouris, L. 2008, 'At the potter's wheel: An argument for material agency', in C. Knappett and L. Malafouris (eds), *Material Agency: Towards a Non-Anthropocentric Approach*. New York: Springer.

Menary, R. (ed.) 2010, *The Extended Mind*. Cambridge, MA: MIT Press/Bradford.

'Military unveils bionic super-soldiers capable of withstanding mental toll of war' 2013, *The Onion*, 9 October, viewed 12 January 2014, http://www.theonion.com/articles/military-unveils-bionic-supersoldiers-capable-of-w,34156.

Office of the Chief Scientist of the Air Force 2010, *Report on Technology Horizons: A Vision for Air Force Science and Technology During 2010–2030*, viewed 13 February 2014, http://www.flightglobal.com/assets/getasset. aspx?ItemID=35525.

Pow, H. 2012, 'More U.S. troops committing suicide than being killed fighting in Afghanistan in "tough year" for armed services', *UK Daily Mail*, viewed 5 January 2014, http://www.dailymail.co.uk/news/article-2222674/More-U-S-troops-committing-suicide-killed-fighting-Afghanistan-tough-year-armed-services.html.

Preston, B. 2013, *A Philosophy of Material Culture: Action, Function and Mind*, New York and London: Routledge.

Rowlands, M. 2006, 'The normativity of action', *Philosophical Psychology*, 19(3): 401–16.

Schaeffer, J-M. 2009, *El fin de la excepción humana*. Mexico, DF: Fondo de Cultura Económica.

Simon, J. 2014, *Distributed Epistemic Responsibility in a Hyperconnected Era*, European Commission, viewed 6 March 2014, https://ec.europa.eu/digital-agenda/sites/digital-agenda/files/Contribution_Judith_Simon.pdf.

Sparrow, R. 2009, 'Predators or plowshares? Arms control of robotic weapons', *IEEE Technology and Society*, 28(1): 25–9.

Thomasson, A. 2007, 'Artifacts and human concepts', in E. Margolis and S. Laurence (eds), *Creations of the Mind: Essays on Artifacts and their Representation*. Oxford: Oxford University Press.

Verbeek, P-P. 2005, *What Things Do: Philosophical Reflections on Technology, Agency, and Design*, University Park, PA: Pennsylvania State University Press.

Verbeek, P-P. 2006, 'Materializing morality: Design ethics and technological mediation', *Science, Technology & Human Values*, 31(3): 361–80.

Verbeek, P-P. 2008, 'Obstetric ultrasound and the technological mediation of morality: A postphenomenological analysis', *Human Studies*, 31(1): 11–26.

Waldby, C. 2000, *The Visible Human Project: Informatic Bodies and Posthuman Medicine*. New York and London: Routledge.

Walker, G., Stanton, N., Salmon, P. and Jenkins, D. 2007, *A Review of Sociotechnical Systems Theory: A Classic Concept for New Command and Control Paradigms*, UK, Human Factors Integration Defence Technology Centre, viewed 4 June 2014, http://www.hfidtc.com/research/command/c-and-c-reports/phase-2/HFIDTC-2–1–1–1-2-command-paradigms.pdf.

Chapter 3

On Human and Machine:
When is a Soldier not a Soldier?[1]

Joseph Pugliese

Inscribed in the titular question of this chapter – When is a soldier not a soldier? – is the tacit understanding that there is a clearly defined boundary line that effectively and clearly demarcates the difference between human and machine. The very category of the super soldier, I contend, would be unintelligible without this presupposition. It is precisely this presupposition that I wish to problematise in the course of this chapter. Rather than attempting to disarticulate the category of the human from the machine in order to find some point of pure and uncontaminated 'humanness', an exercise, as I argue below, that is ultimately untenable, in this chapter I propose an approach that pivots precisely on the inextricable relation between bodies and technologies, humans and machines. In deploying an approach orientated by the twin concepts of *somatechnics* and *prosthetics* (both terms are explained in detail below), my focus will be on theorising the super soldier in terms of a figure produced by a networked assemblage of complex forces and relations.

Constructing the Super Soldier

Across much of the relevant literature, the figure of the super soldier emerges through the discourse of technological and biological *enhancement*: founded upon an undefined yet presupposed baseline of 'humanness', the super soldier is what is constructed through a series of technological interventions and manipulations that transmute the soldier into an enhanced human-machine of war. Precisely because the question as to what constitutes the human is never broached, the literature on super soldiers proceeds as if this were a question that need not be addressed. Only through this critical elision can one begin to speak of 'enhancement'. The concept of enhancement of the human is predicated on an understanding that assumes that there is something intrinsically human about the human that, at some primordial point of origin, is not already enhanced by any technological intervention. In this understanding, enhancement is posited on a notion of the human that remains essentialised, self-identical and *a priori*.

1   My thanks to Constance Owen for her brilliant research assistance.

Because the humanness of the human is never posed or examined, and because the category of the human is made to operate, by implicit definition, as what stands in contradistinction to technology, the literature on the super soldier proceeds to delineate the multiple quandaries that the concept of enhancement throws up when one is discussing the super soldier. As a way of illustrating these quandaries, I want to work through a report by Fritz Allhoff et al. (2009) titled the 'Ethics of Human Enhancement: 25 Questions and Answers', prepared for the US National Science Foundation. Allhoff et al. (2009, p. 5) open their report with a sweeping historical vision: 'Since the beginning of history, we have also wanted to become more than human, to become *Homo superior*. From the godlike command of Gilgamesh, to the lofty ambition of Icarus ... throughout the world's history, we have dreamt – and still dream – of transforming ourselves to overcome our all-too-human limitations'. Allhoff et al. (2009, p. 5) proceed to argue that this epic vision is essential if we are to appreciate where we are today – where 'something seems to be different': 'With ongoing work to unravel the mysteries of our minds and bodies, coupled with the art and science of emerging technologies, we are near the start of the Human Enhancement Revolution'. However, before the authors can proceed to delineate the dimensions of this revolution, they are compelled to articulate a number of fundamental 'Definitions and Distinctions' that accrue around the very concept that underpins the entirety of their report. Allhoff et al. (2009, p. 8) ask: 'What is human enhancement?'

If there is the unspoken assumption in Allhoff et al.'s work that there is a categorical distinction between human and technology, then this distinction is further amplified in their report by the grounding of their discussion of human enhancement on yet another unquestioned binary: natural/artificial. Allhoff et al. (2009, p. 8) write:

> [R]eading a book, eating vegetables, doing homework, and exercising may count as enhancing ourselves, though we do not mean the term this way in our discussion here. These so-called 'natural' human enhancements are morally unproblematic to the extent that it is difficult to see why we should not be permitted to improve ourselves through diet, education, physical training, and so on.

In placing 'natural' in scare marks, Allhoff et al. (2009, p. 9) reflexively signal that this term is somehow problematic and, indeed, the distinction between 'natural' and 'artificial' begins to founder as they proceed in their discussion because, as they explain, 'the natural-versus-artificial distinction, as a way to identify human enhancement, may prove most difficult to defend given the vagueness of the term "natural"'. Despite the dangers presented by this 'vagueness', Allhoff et al. (2009, p. 9) continue to rely on this distinction even as they appear to critique it:

> For instance, if we can consider X to be natural if X exists without any human intervention or can be performed without human-engineered artifacts, then eating food (that is merely found but perhaps not farmed) and exercising (e.g., running barefoot but not lifting dumbbells) would still be considered to be

natural, but reading a book no longer qualifies as a natural activity (enhancement or not), since books do not exist without humans.

As I will discuss further below, the quandaries – conceptual and ethical – that these binary distinctions generate in the literature are many and, I would argue, are largely unproductive in actually resolving some of the key ethical questions that are raised by the emergence of the super soldier. In answering the question of when a soldier is not a human soldier, I want to propose a different tack, one that is no longer reliant on human/technology or natural/artificial binarised distinctions. This different approach is one predicated on two key concepts that I will proceed to deploy in order to critique the sort of categorical presuppositions that I have thus far identified. The two key concepts I propose to deploy are: *somatechnics* and *prosthetics*.

Somatechnics

Somatechnics, as Susan Stryker and I (2009, p. 1) write:

> in its combination of Greek roots, evokes ancient Western philosophical traditions, and seems to solicit the critical reexamination of canonical treatments of *techné*, histories of technology and the arts, the role of the body in the production of knowledge, and phenomenological approaches to problems of epistemology, along with the whole range of bio-technical facts and fantasies that have come to be associated with the *cyborg*. The term *somatechnics* troubles and blurs the boundary between embodied subject and technologized object, and thus between the human and the non-human, and the living and the inert, and it asks us to pay attention to where, precisely, a prosthesis stops and a body starts.

Through the deployment of a somatechnic frame, the natural/artificial distinction that Allhoff et al. maintain in differentiating between running barefoot and lifting dumb-bells becomes untenable. Running barefoot is always already a somatechnology of body modification. There is no purely biological body that is outside of technological mediation in the act of running barefoot. The act of running barefoot achieves its cultural intelligibility precisely as a 'natural' act through its encoding in the technology of language and its predication on a metaphysics of nature versus culture. There is nothing inherently natural about running barefoot: it is always already a culturally mediated act in which the natural and the technological, at the most elementary level of the technics of language, cannot be cleanly and definitively separated. Likewise, informing the act of the gathering of food that is 'merely found' is an entire field of cultural mediation that entails acts of linguistic identification (this is food, this is not food), naming (this is a berry) and classification (edible or not edible) and consequent bodily modification (through degustation, assimilation, excretion and so on).

As Stryker and I contend (2009, p. 2):

somatechnics suggests that embodiment cannot be reduced to the merely physical any more than it can dematerialized as a purely discursive phenomenon. Embodiment is always biocultural, always techno-organic, always a practical achievement realized though some concrete means. At its most quotidian, *somatechnics* references the particular ensemble of embodiment practices operative at any given place at any given time, but it also gestures more grandiosely toward an ontological necessity, a general somatechnic imperative that governs the field of our collective being, from our primate past to our post-human present (and, perhaps, our zoömachinic future): we have never existed except in relation to the *techné* of language, divisions of labor, means of shelter and sustenance, and so forth.

In his theorisation of the relation between body and technology, Jacques Derrida (2002, p. 244) articulates the inextricable tie between the natural (*physis*) and the synthetic (*technè*). He emphasises that this relation 'is not an opposition; from the very first there is instrumentalization ... a prosthetic strategy of repetition inhabits the very moment of life. Not only, then, is technics not in opposition to life, it also haunts it from the very beginning' (Derrida 2002, p. 244). From the very beginning, then, the body is always already mediated by a series of technologies such as language and law. The body, for example, is inscribed from the very beginning (at birth) by a series of laws that proceed to determine its legal identity, its gender, its maternity and paternity and so on. The linguistic inscription of the body must be seen, indeed, to constitute the very conditions of possibility for the conceptual marking of the body as 'human' – in other words, we can only comprehend what is human or not human through the categorical distinctions that are enabled through the technics of language.

Approaching the super soldier through the conceptual frame of somatechnics serves to overturn the catalogue of binaries that inscribe canonical understandings of this figure that are predicated on a number of binarised distinctions: human/technology, natural/artificial. In the process, it also renders superfluous a number of the hollow quandaries that haunt the field. For example, having realised that the nature/artificial distinction will inevitably lead them 'to be mired in ... theological issues', Allhoff et al. (2009, p. 10) then proceed to draw on the internal/external distinction: 'assimilating tools into our persons creates an intimate or enhanced connection with our tools that evolves our notion of personal identity, more so than simply owning things (as wearing name-brand clothes might boost one's sense of self)'. As this other distinction between internal/external is also, from a somatchnics perspective, untenable, it is no sooner examined by Allhoff et al. (2009, p. 10) than it is called into question:

> This is not to say that an enhancement-versus-tools distinction is ultimately defensible or not, but only that it does not help an early investigation into the ethics of using such technological innovations – whatever we want to call them – to consider 'enhancement' so broadly that it obscures our intuitive

understanding of the concept and makes everything that gives us an advantage in life into an enhancement.

Allhoff et al. signal here an insight into a somatechnical understanding of the body/ technology relation that challenges and problematises 'intuitive' understandings of the topic that are informed by untenable binarised relations and unquestioned presuppositions. Everything that gives us an advantage in life is an enhancement. There is no problem here in upholding this 'broad' definition. Working from this premise, one can then proceed critically to examine the question of how different forms of linguistic, cultural and technological mediation work to construct different figures – for example, the super soldier – and to raise the relevant ethical questions that accrue around that specific figure. In other words, rather than flattening out differences, a somatechnical approach to the super soldier asks how different modes of modification work to raise ethical questions that effectively resonate across the social field: for example, 'If we engineer a soldier who can resist torture, would it still be wrong to torture this person with the usual methods?' (Lin 2012). The critical question embedded here is not where the human and the technological begins or ends; rather, the question is: Is it ethical to torture such a subject?

The body of the soldier must be seen as always already mediated by an inextricable relation between corporeality and technology. The very figure of the soldier *as soldier* emerges etymologically from the technology of money: the Latin word *solidus*, signifying a gold coin in the time of the emperors, informs the contemporary understanding of a soldier as a subject that is in the paid employ of a military organisation or state (Lewis and Short 1966, p. 1719). I mark the etymology of the word 'soldier' not as some quaint exercise into the obscure origin of words, but in order to bring into focus how the technology of language works to efface its metaphorical origins (for example, *solidus* as the technology of money = human soldier) through the operations of *catachresis* or dead metaphors. As Umberto Eco (1979, p. 109) remarks: 'When the metaphor becomes customary, a *catachresis* takes place' – that is, the metaphorical origins of the word are forgotten and effaced and consequently supplanted by a literal understanding of the word or concept. From the beginning, inbuilt in the very concept and embodied figure of the soldier is a crucial relation between instrumentalising technologies of money and arms that constitute the term's very conditions of cultural intelligibility. Technology does not arrive later, after the fact of having identified what a soldier is; rather, the very term 'soldier' is animated from its moment of origin by technology, specifically, the technology of money. The concept of the soldier is, when viewed in this light, shot through with a Nietzschean range of military and numismatic resonances: it emerges as a product of a 'mobile army of metaphors … [that] after long use seem solid, canonical, and binding' (Nietzsche 1989, p. 250). Drawing precisely on the metaphor of coinage, Nietzsche (1989, p. 250) describes seemingly unmediated and 'literal' concepts as 'worn-out metaphors without sensory impact, coins which have lost their image and now can be used only as metal, and no longer coins'. In the literature on the super soldier and enhancement, the figure of the soldier operates precisely as a worn-out metaphor that has effaced its original dependence on the role of coinage in order to emerge as a purely human entity uncontaminated by technology, specifically, by the *solidus*.

A somatechnic approach to the figure of the super soldier would effectively bypass the innumerable binaries that inevitably lead to irresolvable contradictions: Where does the human end and the technological begin? What is exterior to the body of the soldier and what interior? Are various forms of augmentation either therapy or enhancement? The attempt to resolve this last question culminates in Allhoff et al. (2009, p. 13) contemplating the 'The Paradox of the Heap':

> Given a heap of sand with N numbers of grains of sand, if we remove one grain of sand, we are still left with a heap of sand (that now has N-1 grains of sand). If we remove one more grain, we are again left with a heap of sand (that now has N-2 grains). If we extend this line of reasoning and continue to remove grains of sand, we see that there is no clear point P where we can definitely say that a heap of sand exists on one side of P, but less than a heap exists on the other side. In other words, there is no clear distinction between a heap of sand and a less-than-a-heap or even no sand at all.

The Paradox of the Heap, in its concern with the impossible task of definitively locating that original point of difference that would effectively deliver a pure concept (for example, 'the human' or 'the technological'), evidences the complex logic that haunts and undoes all essentialist categories. A somatechnical approach works to situate P within the contingency of all the entangled factors that are constitutive of its self-identity without embarking on the impossible task of locating some pure and unmediated point of origin. Situated in this context, it is telling that, in his condemning of transhuman enhancement because it risks destroying the essence of what is human, Francis Fukuyama (2002, p. 149) resorts to defining human essence by drawing on an *undefined* 'Factor X'. Factor X, I contend, is the disavowed technological ghost that haunts Fukyama's fantasy of the unmediated human.

Prosthetics

Building on this somatechnical understanding of the soldier, super or otherwise, I now want to turn to the concept of prosthetics. The Pentagon's high-tech Defence Advanced Research Projects Agency (DARPA) is well on the way to developing a number of prosthetic projects aimed at producing super soldiers. They include: 'Exoskeletons [that] could multiply the strength of soldiers, enabling them to run for hours and carry weights far beyond what is possible unassisted. Contact lenses [that] will transmit images from satellites and drones to soldiers on the battlefield. Helmets [that] could communicate telepathically' (Gayle 2012); and a Warrior Web undersuit 'that would help reduce injuries and fatigue and improve soldiers' ability efficiently to perform their mission' (DARPA 2013). These prosthetic developments that are designed to produce the super soldier must not be seen as mere 'add ons' to a purely human subject that arrives into the military in a type of pure and unmediated state of 'humanness'. Rather, they must be seen as prosthetic augmentations of a figure, the soldier, who is already inscribed by various technological mediations. Operative here is an articulation between seemingly separate parts – the human agent and the technological equipment – that is simultaneously predicated on technically augmenting the power and reach of the human agent through the figure of the prosthetic. Bernard Stiegler (1998, p. 146) suggests that prosthetics is 'a putting outside-the-self that is also a putting-out-of-range-of-oneself'. The exoskeleton, contact lenses linked to satellite and drones, and telepathic helmets are all inscribed by the logic of the prosthetic. The prosthetic enables a putting outside-the-self that is also the putting-out-of-range-of-oneself.

The inextricable relation between humans and technology that I have thus far mapped enables a prosthetic grafting that defies categorical separation of different entities and figures. As I have discussed elsewhere (Pugliese 2013, p. 204), this process of prosthetic grafting is something that is brought into clear focus by the relation between drone pilots and their charges. US Air Force Colonel Matt Martin remarks that, even as he is located in the Ground Control Station of Nellis Air Force Base, Nevada, he views himself as having become completely coextensive with the drone he is piloting, regardless of the fact that the drone is actually flying thousands of miles away in Afghanistan: 'I was already starting to refer to the Predator and myself as "I", even though the airplane was thousands of miles away' (cited in Pugliese 2013, p. 204).

The prosthetic relation between technological platforms and soldiers can be seen to constitute the very figure that Donna Haraway (1991, p. 177) termed the 'cyborg':

> To recapitulate certain dualisms have been persistent in Western traditions; they have all been systemic to the logics and practices of domination of women, people of colour, nature, workers, animals – in short, domination of all constituted others … Chief among these troubling dualisms are self/other, mind/body, culture/nature … High-tech culture challenges these dualisms in intriguing ways. It is not clear who makes and who is made in the relation between human and machine. It is not clear what is mind and what body in machines that resolve into coding practices.

In the prosthetic figure of the super soldier, it is not clear 'who makes and who is made in the relation between human and machine'. The graft of the prosthetic blurs this boundary. Moreover, in the context of the digitised codes that interlink high-tech contact lens, humans, satellites and drones in their practical operations, 'It is not clear what is mind and what is body in machines that resolve into coding practices'. In keeping with the cyborg logic of the prosthetic, there is no 'proper' body in contradistinction to the machine. Perhaps this will be most clearly evidenced by my discussion of DARPA's JASON project.

The JASON project entails the collecting of DNA from military personnel in order to identify genome sequences that define 'ideal' soldiers. These genome sequences, the JASON project (2010, p. 1) suggests, can be used in order genetically to produce super soldiers: 'both offensive and defensive military operations may be impacted by the applications of personal genomics technologies through the enhancement of the health, readiness, and performance of military personnel'. The JASON project (2010, p. 4) is specifically seeking to establish correlations between a soldier's genotype and phenotype in order to produce its 'ideal' super soldier: 'Many phenotypes of relevance to the DoD are likely to have a strong genetic component, for which better understanding may lead to improved military capabilities'. Inscribed here, in this focus on ideal phenotypes, is a long and troubled history of racial science undergirded by hierarchies governed by white supremacist ideologies that I am precluded from discussing due to space constraints (for a discussion of the racial dimensions of genetics, see Pugliese 1999). I do, however, want to discuss the prosthetic dimensions of DNA in relation to the proposed super soldier with her genetically 'improved military capabilities'. The grafting of a genome sequence, identified as crucial in the production of a super soldier, onto the DNA of a soldier does more than raise a cluster of ethical questions; it also brings into focus the manner in which genetic manipulation effectively undoes any clear and definitive division between 'the human' and 'the technological'.

The proposed genomic production of the super soldier, precisely because it so clearly collapses the range of binaries – nature/artifice, subject/object, body/technology – that continue to inscribe discussions in the field, brings into focus another critical dimension that remains to be addressed:

> *somatechnics* suggests the possibility of radically different ways of relating embodied subjectivity to the environment, ways that require a metaphysics not predicated on the subject/object split. Refusing to cut up the world according to this familiar dichotomy, *somatechnics* demands, too, a re-evaluation and reframing of ethics – of the proper regard for the interrelationship between other, self, and world. It raises anew the hoary questions of agency and instrumental will, of freedom and determination (Pugliese and Stryker 2009, p. 1).

A somatechnic reading of the super soldier problematises the binaries of human/technology, subject/object and agent/tool. As I have thus far argued, a critical examination reveals how conceptually untenable these binaries are. This somatechnic understanding of the entangled relation between humans and technology, I suggest, will prepare us to address the envisaged production of super soldiers that will take the form of lethal autonomous robots (LARs) animated by an 'artificial conscience' that invests them with the capacity for cognitive reflection and ethical decision-making:

Now a small group of scholars is grappling with what some believe could be the next generation of weaponry: lethal autonomous robots. At the center of the debate is Ronald C. Arkin, Georgia Tech professor who has hypothesized lethal weapons systems that are ethically superior to human soldiers on the battlefield. A professor of robotics and ethics, he has devised algorithms for an 'ethical governor' that he says could one day guide an aerial drone or ground robot to either shoot or hold its fire in accordance with internationally agreed-upon rules of war (Troop 2012).

The prospect of super soldiers cast in the form of lethal autonomous robots animated by both 'artificial' intelligence and 'artificial' conscience or 'moral intelligence' is no mere science fiction fantasy (see Arkin 2009, pp. 38–9; Wallach and Allen 2009, pp. 66–8). Even as I write, the United Nations is hosting a 'Meeting of Experts' in Geneva, Switzerland, to discuss the critical issues that will be generated by the projected rise of lethal autonomous robots in the conduct of war (International Committee for Robot Arms Control 2014). What interests me, from the perspective of the problematics that I have so for mapped in making categorical divisions between human soldiers and technology, is that the very category of LARs is one that is also proving difficult to define. This definitional indeterminacy is exemplified by the fact that the Autonomous Robotics Thrust Group of the Consortium on Emerging Technologies, Military Operations, and National Security has raised 'serious questions about what an LAR actually is. Certainly, it has a technology component, but in some ways this is almost trivial compared to its social, political, ethical and cultural dimensions' (Marchant et al. 2011, p. 287). Underscoring the importance of this constellation of social, political and ethical factors, the group declares: 'In fact, one might well argue that in many important ways a LAR is more of a cultural construct than a technology' (Marchant et al. 2011, p. 287). I understand the group to be saying here that a LAR cannot, as a technology, be fully understood outside of all the complex cultural factors that constitute its conditions of intelligibility. As I have been contending throughout the course of this chapter, the insight articulated by this group, that goes so far as to classify the technology component of LARs as 'trivial' in comparison to the set of all the other factors, resonates for me in terms of the complex forces and relations that are constitutive of the super soldier.

Conclusion

In approaching my conclusion, I want to situate the question – When is a soldier not a soldier? – within the complex matrix of bodies, technologies, non-biological substrates and ground that effectively blurs the lines between one and the other in order to underscore the impossibility of addressing this question in a categorical manner. In the context of this matrix of entangled entities, I do not invoke 'ground' in a purely metaphorical manner. On the contrary, I gesture to the impossibility of articulating an understanding of something as 'natural' as 'earth' that is not always already inscribed by some form of technology. I refer here, for example, to the emergence of the 'intelligent landscape', as produced by the Israeli-based G-Max Security, a company that is enmeshing sensor and surveillance technologies and geotextiles within fields and forests. 'This security geotextile', writes Geoff Manaugh (2012), 'is, in effect, an electromagnetic nervous system in the ground'. In his discussion of intelligent landscapes, Manaugh traces the many militarised uses to which they can be put. The buried sensor and surveillance technologies, which can be covered by grass and trees, can be interlinked to satellites and drone Ground Control Stations in order to produce a 'semi-autonomous landscape-to-robot constellation' (Manaugh 2012). This intelligent landscape could be used both to guide drones for precision landings on geotextile fields or, alternatively, to jam and repel foreign drones or manned aircraft from one's sovereign airspace. 'What you think is a forest is a complex signaling landscape. What appears to be a garden is a computational geotextile interacting with driverless ground vehicles miles away' (Manaugh 2012). Operative here are a range of non-human *actors* instrumental in shaping, conditioning and producing new entities (for example, landscape-to-robot constellations) and cultural practices (for example, military precision landings); and I use 'actor' here in Bruno Latour's (2004, p. 226) extended sense: 'things might authorize, allow, afford, encourage, permit, suggest, influence, block, render possible, forbid and so on, in addition to "determining" and serving as a backdrop to human action'. Complicating simplistic understandings of cause and effect, Latour (2004, p. 226) locates agency in nonhuman objects and things: 'anything that modifies a state of affairs by making a difference is an actor'.

Situating the super soldier in an intelligent landscape works to evidence the manner in which this figure is inextricably enmeshed in and enabled by a *networked somatechnic assemblage* of actors that includes bodies, technologies and geotextile ground. The super soldier is the 'effect' of this networked assemblage of actors and its complex relays of power. This networked assemblage must, of course, be further situated within the larger ensemble of the military-industrial-complex and its ongoing expansive colonisation and militarisation of civic spaces (see Pugliese 2013, p. 190) and civilian technologies – for example, DARPA is working on the drone cooptation of such civilian communication interfaces as 'Skype and developing and implementing rich user interfaces to display what is happening in a sensor array on a Google-Maps-like interface' (InformationWeek 2011).

In the wake of their tracking of the complex issues that inscribe the field of human enhancement, Nick Bostrom and Julian Savulescu (2013, p. 3) suggest that the most viable and fruitful approach would be one that 'would reflect the concrete circumstances and consequences of particular enhancement practices: Precisely what capacity is being enhanced in what ways? Who has access? Who makes the decisions? Within what cultural and sociopolitical context?' They conclude that 'Justifiable ethical verdicts may only be attainable following a specification of these and other similarly contextual variables' (Bostrom and Savulescu 2013, p. 3).

Positioned within this field of contextual variables, the super soldier cannot be seen as operating as an autonomous entity; neither can s/he be seen as an entity that can be categorised through the deployment of a neat conceptual line of demarcation between human and machine. On the contrary, s/he is critically constituted by a networked assemblage of complex forces and materialities that bring into question traditional dichotomies and that demand a critical re-evaluation of ethics, laws of war and questions concerning identity, agency, freedom and determination.

References

Allhoff, F., Lin, P., Moor, J. and Weckert, J. 2009, 'Ethics of Human Enhancement: 25 Questions and Answers', U.S. National Science Foundation, viewed 21 August 2013, http://ethics.calpoly.edu/NSF_report.pdf.

Arkin, R.C. 2009, *Governing Lethal Behavior in Autonomous Robots*. Boca Raton, FL: CRC Press.

Bostrom, N. and Savulescu, J. 2013, 'Introduction: Human Enhancement Ethics: The State of the Debate', in S. Savulescu and N. Bostrom (eds), *Human Enhancement*, Oxford: Oxford University Press.

DARPA 2013, 'Warrior Web closer to making its performance-improving suit a reality', DARPA News Events Releases, 22 August 2013, viewed 3 October 2013, http://www.darpa.mil/NewsEvents/Releases/2013/08/22.aspx.

Derrida, J. 2002, *Negotiations: Interventions and Interviews 1971–2001*. Stanford, CA: Stanford University Press.

Eco, U. 1979, *A Theory of Semiotics*. Bloomington, IN: Indiana University Press.

Francis F. 2002, *Our Posthuman Future: Consequences of the Biotechnology Revolution*. New York: Picador.

Gayle G. 2012, 'Army of the Future: Soldiers will be able to run at Olympic speed and won't need food or sleep with gene technology', *Daily Mail*, 12 August, viewed 22 July 2014, http://www.dailymail.co.uk/sciencetech/article-2187276/U-S-Army-Soldiers-able-run-Olympic-speed-wont-need-food-sleep-gene-technology.html.

Haraway, D. 1991, *Simians, Cyborgs, and Women: The Reinvention of Nature*. New York: Routledge.

InformationWeek 2011, 'DARPA eyes mobile apps to fly drones', 6 December 2011, viewed 15 May 2014, http://www.informationweek.com/mobile/darpa-eyes-mobile-apps-to-fly-drones/d/d-id/1101707.

International Committee for Robot Arms Control 2014, 'ICRAC statement on technical issues to the UN CCW Expert Meeting', 14 May 2014, viewed 15 May 2014, http://icrac.net/2014/05/icrac-statement-on-technical-issues-to-the-un-ccw-expert-meeting.

JASON 2010, 'The $100 Genome: Implications for the DoD', JASON, The MITRE Corporation, Virginia, viewed 9 May 2014, http://www.fas.org/irp/agency/dod/jason/hundred.pdf.

Latour, B. 2004, 'Nonhumans', in S. Harrison, S. Pile and N. Thrift (eds), *Patterned Ground: Entanglements of Nature and Culture*. London: Reaktion Books.

Lewis, C.T. and Short, C. 1966, *A Latin Dictionary*. Oxford: The Clarendon Press.

Lin, P. 2012, 'More than human? The ethics of biologically enhancing soldiers', *The Atlantic*, 16 February, viewed 3 March 2014, http://m.theatlantic.com/technology/archive/2012/02/more-than-human-the-ethics-of-biologically-enhancing-soldiers/253217.

Lin, P., Maxwell, J.M. and Abney, K. 2013, 'Enhanced Warfighters: Risks, Ethics, and Policy, 1 January 2013', pdf downloaded 20 February 2013, http://www.ethics.calpoly.edu/Greenwall_report.pdf.

Manaugh, G. 2012, 'Drone landscapes, intelligent geotextiles, geographic countermeasures', BLDG BLOG, 7 January 2012, viewed 1 September 2012, http://bldgblog.blogsplot.com/2012/01/drone-landscapes-intelligent.html.

Marchant, G.E., Allenby, B., Arkin, R., Barrett, E.T., Borenstein, J., Gaudet, L.M., Kittrie, O., Lin, P., Lucas, G.R., O'Meara, R. and Silberman, J. 2011, 'International Governance of Autonomous Military Robots', *The Columbia Science and Technology Law Review*, XII: 272–315.

Nietzsche, F. 1989, *Friedrich Nietzsche on Rhetoric and Language*, S.L. Gilman, C. Blair and D.J. Parent, eds and trans., New York and Oxford: Oxford University Press.

Pugliese, J. and Stryker, S. 2009, 'Introduction: The somatechnics of race and whiteness', *Social Semiotics*, 19(1): 1.

Pugliese, J. 1999, 'Identity in question: A grammatology of DNA and forensic genetics', *International Journal for the Semiotics of Law*, 12(4): 419–44.

Pugliese, J. 2013, *State Violence and the Execution of Law: Biopolitical Caesurae of Torture, Black Sites, Drones*. Abingdon and New York: Routledge.

Stiegler, B. 1998, *Technics and Time, 1: The Fault of Epimetheus*. Stanford, CA: Stanford University Press.

Troop D. 2012, 'Robots at War: Scholars Debate the Ethical Issues,' *The Chronicle of Higher Education*, 10 September, viewed 26 October 2013, http://chronicle.com/article/Moral-Robots-the-Future-of/134240.

Wallach, W. and Allen, C. 2009, *Moral Machines: Teaching Robots Right from Wrong*. Oxford: Oxford University Press.

Chapter 4

The Super Soldier as Scholar: Cultural Knowledge as Power

Barbara Gurgel and Avery Plaw

The quest for better soldiers is as old as the human preoccupation with war, which dates from long before humans began to record their history (Keeley 1996). Virtually every civilisation has had a caste of elite soldiers, from the Zande of Central Africa to the British Special Air Service. Military leaders have experimented with diverse methods of molding these elite soldiers. The Spartans had the 13-year Agoge, inhumane but successful; the Samurai had the Bushido and began martial arts training at the age of five; and the Navy Seals have their famously gruelling two years of training. In addition to training, militaries in general, and the US military in particular, have long sought superior equipment, and especially superior weapons. President Obama dramatised this project when he recently declared, 'we are building Iron Man' (Matyszczyk 2014). It is this complementary integration of superior training and superior equipment that explains the origin of 'super soldiers'. While many of the chapters in this volume focus on the technology side of 'soldier enhancement', this chapter examines a new frontier in training, and one that is of special importance for a global power like the United States whose forces could be suddenly sent anywhere in the world: cultural training.

Cultural training is vitally important if American service members sent abroad are to function with optimal effectiveness in foreign environments – a point all too well illustrated by controversies surrounding incidents of cultural insensitivity, such as the controversies over discarding and abusing Qur'ans (Rahimi and Rubin 2012; Cocks 2008), peeing on the dead enemy (Rosenberg 2012), humiliating local women (Bordin 2011, p. 12) and a long list of other unfortunate and counter-productive episodes. In an era in which major military operations are increasingly concerned with counterinsurgency (COIN) and the winning of hearts and minds, cultural understanding is crucial (Center for Army Lessons Learned [CALL] 2009). Unfortunately, the pattern of preparation remains that service members receive, at best, 'cliff notes' – like training and a packet of handy phrases (Lewis 2006, p. 3) before being immersed in societies which are entirely foreign to them. Cultural 'blunders' have always been a part of deployments, but more and more it is becoming clear that an inability to identify with the people of a foreign culture can keep service members from carrying out their missions to the best of their abilities. The creation of an 'Iron Man' is all well and good, but if the service member in the suit has only a tenuous grasp on the cultures in which s/he will be operating, and proceeds to terrify, insult and repel the locals, that service member will be facing a real handicap. Combat training and technology can only go so far.

In this chapter we will first outline the current challenges facing the US military, pointing to incidents which exemplify the difficulties of communicating across cultures. Second, we will describe the US military's recent efforts to address these challenges, disparate and event-specific as they are. Finally, we will review several reports from experts in these fields, and draw some preliminary conclusions about what has been, and needs to be, done.

Challenges in Cross-Cultural Communication

Examples of reckless disregard for local cultural norms by American service members are plentiful in places like Japan (Spitzer 2012) and Europe (Abbe and Gouge 2012), but they are recently most prevalent in the Middle East, in particular in Iraq and Afghanistan. Reports of abusing the Qur'an are among the most common, along with posing with or urinating on the bodies of the enemy fallen, blatantly disrespecting tribal leaders, male service members behaving inappropriately towards women or even serving an endless supply of pork in the cafeteria shared by Western and Afghan soldiers and officials (Cocks 2008; Rosenberg 2012; Nordland 2012). These are all things that American service members should not need special cultural training to know is disrespectful in Iraq and Afghanistan. And yet, these incidents continue to occur.

Even service members who try to be conscious of cultural differences make cultural mistakes, through lack of experience and training. Common lesser-known cultural blunders include things like wearing sunglasses when speaking to tribal elders (where hiding your eyes while speaking is seen as suspicious or insincere), walking or standing in front of someone who is praying (where this is a sin and an insult), placing boots or shoes on a table or showing someone the bottoms of your feet (in cultures in which this indicates that that person is beneath you), blowing your nose in public and locker room-like nudity among service members (never done in Middle-Eastern culture; TRADOC 2006, pp. 25, 35; Brown 2008, p. 445). Even appearing in a hurry when speaking to civilians (Rosenberg 2012) or asking a male about female family members (Brochure 2012) are perceived as rude in many local cultures. These types of accidental cultural miscues can carry the same consequences as the previously mentioned cultural provocations, alienating civilians and breeding tension between Western and Afghan soldiers (CALL 2009). These potential effects are well illustrated in a cultural guide published for Afghan forces which aims to help Afghan soldiers working with coalition forces understand some of these unintentionally insulting cultural differences. The 28-page document included things like these:

- 'If a coalition solider is excited or wants to show appreciation for your work he may pat you on the back or shoulder. It is not meant as an insult and you should not take it personally'.
- 'As you know, Afghans don't blow their noses in the presence of others. But the practice of blowing your nose in public is a very common practice among the countries where your coalition partners come from'.
- 'As you know, Afghans never shake with their left hands, wink, signal with their fingers, or show their private parts in the presence of others in the same shower. But coalition forces have a different way of doing things. They don't want to insult you; it is only a cultural difference. You should talk with your coalition colleagues about these differences' (Brochure 2012).

However, with the rampant illiteracy of Afghan forces, it is not likely that this guide or others like it are the best way to educate Afghan soldiers on the cultural differences of coalition forces.

These sorts of social mistakes may seem of secondary importance in the midst of a war, but cultural training is about more than not alienating the population in which service members are operating. The inability to communicate across the cultural gap costs lives. For example, between 2006 and 2007, American service members killed or wounded 429 Iraqi civilians at checkpoints (Youssef 2007). At these checkpoints, American service members used hand gestures to signal civilians to stop or go, fired warning shots if the signals were ignored and although there were supposed to be other steps in this hierarchy, American service members were often described as jumping quickly to the final action: shoot to kill (Youssef 2007). At least part of the problem at checkpoints in 2006 seemed to be the inconsistent and confusing use of Western, Military and Iraqi hand gestures for 'stop' and 'go' (Bender 2007). In Western societies, an outstretched hand with the palm facing forward means 'stop'. In Iraq, and in Afghanistan, the same signal means 'welcome' (Youssef 2007). The result was a confusion which was too often deadly as American service members opened fire on civilians who appeared to ignore their signals.

Cultural misunderstandings do not only result in harm to local civilians. Between 2007 and 2011, 6 per cent of all hostile coalition deaths were Western service members killed by Afghan soldiers (Bordin 2011, p. 3). Although those types of attacks on Western service members by Afghan soldiers are routinely dismissed as incidents of infiltration, an unclassified 2011 coalition report titled 'A Crisis of Trust and Cultural Incompatibility' shows that a substantial number of these incidents can be attributed to the lack of cultural understanding between the two forces. Clearly, the issue of cultural competence is about more than just hurt feelings (Bordin 2011).

US Efforts to Overcome Cultural Insensitivity

What can be done about these harmful incidents of cultural insensitivity? This section examines efforts by US Armed Forces to put an end to them. Although the US is by no means the only country whose Armed Forces struggle with cultural competence, its strength, its global reach and its history of struggling with this issue make it an exemplary case when discussing issues of culture in conflict.

Some branches of the American military do actually have a long history of recognising the importance of culture in combat. The Special Forces and the Marines have for many years sought to 'culturalize warriors' (Brown 2008, p. 444) and, unlike their counterparts in other branches of the military, seem to understand culture's vital importance to the mission. In his 2008 report '"All They Understand is Force': Debating Culture in Operation Iraqi Freedom', Keith Brown focuses on describing some of the methods various branches within the US military employ to bridge the cultural gaps between themselves and the civilians and allied forces surrounding them. Brown writes that the difference between the Army Special Forces and other sister services lies in the culture of the Special Forces (SF), and that the roots of this difference go back to SF's original mission as established by President Kennedy to 'increase US capacity in counterinsurgency' (Brown 2008, p. 449). He suggests that their often-criticised willingness to 'go native', to adopt local dress and customs, is in fact a reflection of their understanding of the importance of culture in combat – for example, the Special Forces were encouraged to grow moustaches and facial hair to build rapport with the local population during the Gulf War (Brown 2008, p. 451).

Brown also argues that the Marines have shown a history of cultural awareness since Vietnam, and that in Iraq, they 'led the push for new tactics that would replace heavy-handedness with "patience and subtlety"' (Brown 2008, p. 445). The Marines also began 'two differently oriented initiatives in the cultural field': the Center for Cultural Intelligence, and the Center for Advanced Operational Culture Learning (CAOCL), which, from 2006 trained units deployed to Iraq, and now offers cultural training courses on their website, which is required for non-commissioned officers, warrant officers and officers, but which is available for anyone to access. The Marines are even credited with the creation of a working hierarchy of definitions of cultural awareness, from the lowest form, 'cultural consideration', to the highest, 'cultural competence' – meaning 'the ability to utilize knowledge skills, abilities, and attributes to more effectively interact in a socially complex environment' (Grant and Farrell 2013).

In addition to the serious efforts made by Special Forces and the Marine Corps, almost everybody else in the American military seems to at least make some nod (though much more limited) to cultural training. Each branch has separate programs which employ different methods. The best-known example is the US Army Training and Doctrine Command (TRADOC), established in 1973, which is the body in charge of the training regimens of the US Army. It oversees 32 different schools with different areas of expertise, such as the Defense Language

Institute, the School of Advanced Military Studies and the University of Foreign Military and Cultural Studies. Everybody in the army has to go through at least one of these schools. As these examples suggest, at least three schools are concerned with understanding and operating in foreign cultures.

Aside from the language training and other similar types of training that only some service members receive, the only cultural training that most service members get is restricted to fact sheets, pamphlets and briefings before deployments (Nordland 2012). Service members receive 'culture smart cards' with a tourist-like list of 'dos' and 'don'ts' and handy phrases (Brown 2008, p. 445).

In fairness, the TRADOC fact sheets do convey useful information. Sections on body language, suggestions for successful interactions and on appropriate behaviour in mosques and private homes include cultural information that is vital to the success of service members deployed to the Middle East. However, the handbook is designed, by TRADOC's own definition, as an informal 'hip pocket training resource' (TRADOC 2006, p. ii). Any more nuanced insights into that culture are left for service members to learn on their own, taking lessons from their higher ranking and previously deployed counterparts. Captain Thompson, a soldier interviewed for this study who did tours in both Iraq and Afghanistan, explained that part of the jobs of senior service members was basically to 'make sure the "new guys" don't "mess up"', but that service members were largely left to their own devices to really understand the local culture (Thompson 2014).

TRADOC releases not only fact sheets but also pamphlets. One pamphlet from 2008, for example, is titled *The US Army Study of the Human Dimension in the Future: 2015–2024* (TRADOC 2008). The pamphlet explores the challenges of training and maintaining a fighting force that is equipped to operate 'amongst populations with diverse religious, ethnic, and societal values' (TRADOC 2008, iii), and it concedes that 'existing accessions, personnel, and force training and education development efforts will not meet these future challenges' (TRADOC 2008, p. 10). The solution presented in the pamphlet includes things like an increased focus on improvement of capabilities to address these human dimensions (TRADOC 2008, p. 20). According to this pamphlet, 'developing cultural intelligence' among both service members and civilians is the key to success in current and future conflict (TRADOC 2008, p. 205).

Perhaps the best known and most important attempt to address cultural competency in Iraq and Afghanistan, the Human Terrain System, was initiated in 2006 and involved the introduction of civilian social scientists into the battlespace rather than the training of the actual service members. At the time the program was initiated, the US was shifting from a light footprint strategy to classic COIN doctrine in both Iraq and Afghanistan. To be successful a COIN strategy requires cultural acuity – in General Petraeus' words, 'You have to understand not just what we call the military terrain … the high ground and low ground. It's about understanding the human terrain, really understanding it' (quoted in Dehghanpisheh 2008). The point is elaborated in the first chapter of the *Human Terrain Team Handbook*: 'The human dimension is the very essence of irregular warfare environments.

Understanding local cultural, political, social, economic, and religious factors is crucial to successful counter-insurgency and stability operations, and ultimately to success in the war on terror' (Finney 2008, p. 3).

However, because US strategic doctrine had focused since the Cold War on confronting state opponents in conventional battles over territory, 'cultural understanding', in the words of Major Grant Fawcett, 'was never integrated into the doctrine or training of the [US] military' (Fawcett 2009, p. 32). In June 2007 this oversight in army doctrine was corrected, at least nominally, with the publication of Field Manual 3-0 which updated the traditional list of factors that field commanders need to consider in planning operations by adding 'Civil Considerations'. In this way, the traditional commander's mnemonic METT-T (Mission, Enemy, Terrain and weather, friendly Troops and support available – and Time) became METT-TC with the C representing 'Civil Considerations' (Department of the Army 2008, pp. 1–9). But this update raised the question of how commanders could be supplied with a timely 'understanding [of] local cultural, political, social, economic, and religious factors' which they could factor into their calculations (Finney 2008, p. 3).

The task of the Human Terrain System (HTS) was to provide this timely information. In particular, it was designed to 'recruit, train, deploy, and support an embedded operational focused socio-cultural capability; conduct operationally relevant socio-cultural research and analysis; develop and maintain a socio-cultural knowledge base, in order to enable operational decision-making, enhance operational effectiveness and preserve and share socio-cultural knowledge' (Hamilton 2011, p. 1). The first six teams were deployed to Iraq and Afghanistan by 2007. The *Human Terrain Team Handbook* (Finney 2008) describes the teams as follows:

> Human Terrain Teams (HTTs) are five- to nine-person teams deployed by the Human Terrain System (HTS) to support field commanders by filling their cultural knowledge gap in the current operating environment and providing cultural interpretations of events occurring within their area of operations. The team is composed of individuals with social science and operational backgrounds that are deployed with tactical and operational military units to assist in bringing knowledge about the local population into a coherent analytic framework and build relationships with the local power-brokers in order to provide advice and opportunities to Commanders and staffs in the field (Finney 2008, p. 2).

HTTs were deployed at the brigade, division and corps levels (Fawcett 2009, p. 28). Unfortunately, the HTS program generated enormous controversy from its inception. To begin with, it was widely condemned by academic anthropologists and ultimately the American Anthropological Association Executive Board (AAA Executive Board 2007). Partially as a result of this, the program has had enormous difficulty in recruiting sufficient numbers of qualified academics (Fawcett 2009, p. 40; Lamb et al. 2013, pp. 48–51, 120; Stanton 2013, p. 6). There have also been accusations that the training provided to teams was very rushed and shoddy and the result has been field teams that are unprepared for combat zones or to produce valuable analysis, to the degree that they have been a net burden on operations (AAA Commission on the Engagement of Anthropology with the US Security and Intelligence Communities 2009, pp. 46–8; Lamb et al. 2013, pp. 48–50; Stanton 2013, p. 8). Complaints have also arisen concerning the degree of oversight of, and discipline in, teams in the field. At least three team members have been killed in theatre, others gravely wounded and one HTT team member shot and killed a prisoner while he was in custody (Lamb et al. 2013, pp. 53–60; Memorandum 2010, paragraph 3; Stanton 2013, pp. 1, 13, 17, 21). Concerns have also been raised about the number and quality of reports actually produced, and whether the method of reporting to Commanding Officers has been effective (Jebb et al. 2008, p. 7; Lamb et al. 2013, pp. 186–92, 174, 115–17, 51–9; Stanton 2013, pp. 12–13, 35).

None of this is to suggest, however, that the program did not make positive contributions to the US war efforts (as will be shown in the next section, there is at least some evidence that it did). It is, however, to suggest that the record appears to be mixed – that there were and arguably remain significant failures both at the level of the design and realisation of the program. The following briefly reviews some public reports and data on the effectiveness of the HTS and other programs designed to improve cultural sensitivity.

The Effectiveness of Cultural Sensitivity Programs and the Way Forward

In 2010, following the murder of six American soldiers by Afghan Border Police, approval was given to conduct an investigative field study on the cause of green-on-blue violence in Afghanistan. *A Crisis of Trust and Cultural Incompatibility: A Red Team Study of Mutual Perceptions of Afghan National Security Force Personnel and US Soldiers in Understanding and Mitigating the Phenomena of ANSF-Committed Fratricide-Murders* was published a year later. This study systematically interviewed hundreds of Afghan military, police and interpreters, asking questions about their perceptions of American and Western forces. The study also questioned American service members about their perceptions of the Afghan National Security Forces (ANSF). The views on both sides were overwhelmingly negative.

The findings of this study were that, despite the commonly touted explanation that green-on-blue attacks occur because the ANSF were infiltrated by insurgents, a significant fraction of these incidents were actually a result of individual disagreements stemming from cultural differences (Bordin 2011, p. 5). Of the 25 fratricide/murders investigated, 10 of them followed verbal altercations (Bordin 2011, p. 55). In the cases where other ANSF were present at green-on-blue attacks, they often did little to interfere, allowing eight of the shooters in these 15 green-on-blue attacks examined to escape (Bordin 2011, p. 4).

The study found that despite some evidence of improvement in the cultural training of deployed service members, and the fact that the majority of service members felt that they were being culturally aware, the most common complaint on the part of ANSF personnel, and their explanation of the green-on-blue violence, is the widespread cultural insensitivity and ignorance displayed by US soldiers (Bordin 2011, p. 34). ANSF personnel had a long list of complaints, some of which were the consequence of simple differences in manners (for example, service members sitting with their legs uncrossed in front of elders, throwing gifts for children onto the ground, wearing sunglasses during discussions), and some of which reflected a serious gap in the cultural training of deployed US service members (Bordin 2011, pp. 19, 21). The most common complaints were urinating and defecating in public while on patrol (especially in front of women), prolific use of the term 'Mother Fucker' to refer to ANSF personnel and a general arrogance and crudeness that ANSF found unbearable (Bordin 2011, pp. 14–15, 35). Many ANSF personnel freely admitted that the only reason that they have not fought US soldiers is because of explicit orders to the contrary (Bordin 2011, p. 35). In short, the ANSF saw US service members as crude bullies, with little regard for ANSF or Afghan civilian life (Bordin 2011, pp. 12–13).

The study did note several areas where attitudes were more positive or improved. ANSF members tended to view Marines and female service members 'as having better attitudes and being more respectful and respected' (Bordin 2011, pp. 22, 37). Indeed, many of the Afghans interviewed thought most American service members were 'friendly, polite, helpful, brave and very hard workers' (Bordin 2011, p. 23). ANSF members also approved of the use of women to search women and to take the lead in house searches in general (Bordin 2011, p. 23).

Reports on the HTS have also been mixed, but certainly contain some success stories. For example, a 2011 report by the Center on Irregular Warfare and Armed Groups contained several case studies which illustrate positive impacts. The following is one example of a small scale:

> ... a Canadian regiment established a combat outpost in an abandoned home in the center of Nakhonay. After they moved in, they noticed a sharp uptick in the acts of violence directed against them. This violence did not match Human Intelligence (HUMINT) reporting, which indicated that Taliban fighters had left the area to harvest poppies in Helmand. General Menard of TFK instructed HTS members to go into the field to determine the source of the hostility and what could be done to fix it.

> A team of six HTS members visited the area and interviewed locals. The team learned that a sub-tribe was angry at the regiment for having moved into one of the local homes. This home was owned by a village elder, a man from their tribe who lived in Kandahar half of the year and in his Nakhonay home the other half of the year. The Canadian regiment hadn't realized this when they selected the home as their outpost station. The violence, then, was the sub-tribe's method of retaliation; they were attacking the Canadian troops in an effort to drive them from the elder's property. After the HTT discovered this, they organized a

meeting between the Platoon and the elder, ultimately devising a compensation repayment plan covering the regiment's occupation of the compound. Within days, violence in the area noticeably decreased and relations between the community and the Canadians improved (Nigh 2011, p. 26).

Many similar stories are reported (Nigh 2011, pp. 24–5, 28–44; Cox 2011, pp. 27–9). But the question remains whether these constructive contributions from HTTs are having a tangible collective impact.

Perhaps the most rigorous, comprehensive and authoritative report on the HTS to date was published by a team of researchers from the National Defense University in June 2013. In addition to drawing their own conclusions based, among other things, on 105 interviews, this report reviews and integrates the findings of prior studies. Despite recognising manifold problems with the program (and particularly its management), the authors observe that the consistent finding across reports is that 'the large majority of commanders queried found the teams helpful' (2013, pp. 169, 2–3).

For example, HTS Management itself performed two internal studies, one of the initial HTT sent to Afghanistan, one of the five teams initially sent to Iraq. The first report was 'strongly positive', the second more mixed, but both featured very strong endorsements from the Commanding Officers (COs) under whom the HTTs were operating (Lamb et al. 2013, p. 170). This pattern of support from most COs was corroborated by the independent studies which followed (see Table 4.1 below for a summary).

Table 4.1   Comparison of studies sampling commander HTT assessments

Study	Successful	Partial Success	No Impact
West Point Study	High Valued (4)		
CNA Study	Very Useful (5)	Varied Usefulness (8)	Not Useful (3)
IDA Study	Successful (26)	Partial Success (9)	No Impact (1)
NDU Study	Effective (8)	Mixed (4)	Not Effective (1)

Source: Lamb et al. 2013, pp. 175–6.

The first of these was a 2008 report produced by a group of faculty from the United States Military Academy at West Point in which each of the commanders interviewed 'indicated they valued the work of the HTTs' (Jebb et al. 2008, p. 3). The report also stressed, however, that only the teams which were able to convince the commander quickly that they were high-value assets were likely to be really successful (Jebb et al. 2008, p. 3). A Congressionally Directed Assessment of the Human Terrain System was also completed in 2010, and concluded that the program 'has been, in many ways, a success' and 'continues to have strong support from perhaps its most significant constituent: commanders in the field' (Clinton et al. 2010, pp. 2–3). However, it also found that 'real problems exist within the HTS program', notably concerned with training and management (Clinton et al. 2010, pp. 2–3, 60).

Another study, by the Institute for Defense Analysis and completed in 2011, found that the HTT's effectiveness 'may strongly depend upon its ability to develop a functional communication pattern and working relationship with the supported command' and in particular with the commander (Institute for Defense Analysis 2011, pp. 6–5). Fortunately, the majority of the subjects they interviewed (which included 17 brigade commanders) reported constructive relationships, and in fact that the HTT made constructive contributions without which the unit 'could not have been successful' (see table above). Still, many of the interviews included many concerns about the difficulty of communication and general coordination between HTTs and the military hierarchy.

Finally, a TRADOC report released in 2013 found that teams ranged from 'providing significant value' to 'ineffective', but a clear majority of COs reported that their team was the former rather than the latter (Memorandum 2010).

There are also some independent indicators which suggest some mitigation of tension between US forces and their Afghan allies. Most importantly, the green-on-blue violence which was a central motivator of the 2011 Red Team study of 'Cultural Incompatibility' shows significant improvement. For example, the *Long War Journal* (2014, as of 1 October) reports the following numbers:

Table 4.2   Green-on-blue violence

Year	Coalition Soldiers Killed in Green-on-Blue Violence	Total Coalition Fatalities Resulting from Green-on-Blue Violence	Proportion of Coalition Fatalities Resulting from Green-on-Blue Attacks	Coalition Soldiers Injured in Green-on-Blue Incidents
2008	2	2	Less than 1%	3
2009	5	12	2%	11
2010	5	16	2%	1
2011	16	35	6%	34
2012	44	61	15%	81
2013	13	14	9.9%	29
2014	4	4	6%	6

Note: Each of the four categories listed in the table shows an accelerating growth in violence peaking in 2012, and a rapid reduction over the following two years to levels nearing those in 2008 in most respects. This decline appears to begin in the year after the HTT teams reached their full complement (of 30) in Afghanistan in 2011 (see table in Lamb et al. 2013, p. 47).

Of course, there are doubtless a range of factors other than improved cultural sensitivity influencing these numbers, such as reduced cooperative exercises, improved precautions, reductions of coalition forces in Afghanistan and so on. If one were to trace the number of US service members in Afghanistan from 2003 to early 2014 (see table in Washington Post 2014), it can be quickly be identified that the number of green-on-blue fatalities does not closely track the number of US troops in the country – for instance, the number of fatalities almost double from 2011 to 2012 even as troop levels declines (from 100,000 to around 90,000). Moreover, the proportion of the fall in fatalities after 2012 (from 61 to 13 to 4 in the year up to October 2014) is vastly greater than the reduction in troops.

The idea that there is an important relationship between the reduction of cultural tensions (generated through programs like HTS) and reduced green-on-blue violence, although obviously speculative, is strongly supported in the findings of the 2011 Red Team Study. The relationship also finds some reinforcement in the fact that the rapid acceleration and peak of violence corresponds with a series of high-profile incidents of cultural insensitivity (such as the burning of Qur'ans at Bagram Airforce Base on 22 February 2012) and Afghan protests of them, and the decline thereafter has been accompanied by a reduced number of such incidents.

Conclusion

There are at least some indicators, then, that suggest that US efforts to diminish behaviour that is grossly offensive culturally have had some success and some positive impact on coalition operations. However, the programs that have been introduced, most notably the HTS, have also been expensive (with the cost of HTS rising to $160 million in 2010 alone – see Table above) and plagued with setbacks and controversies. Moreover, they are at best diminishing a negative effect (the alienation of locals) rather obtaining a positive effect (constructively advancing US war goals). Yet the very logic that the US military has embraced as a justification for these programs – that is, the centrality of the cultural terrain as a strategic consideration in contemporary warfare – demands more than that we avoid disasters in this area; it requires that we make it a central focus of our planning and preparation for future conflicts, including the way that we train our service members. We need to take positive steps to harness the potential of cultural understanding to more rapidly and effectively achieve our war aims. But how can this be done?

We suggest that the programs of cultural sensitisation described here, however successful they may have been in defusing a potentially disastrous situation, are inadequate in the longer term for two main reasons. The first is that they are too reliant on non-military personnel. The second is that they aim at providing commanders a deeper understanding of their cultural environment, but are content to supply ordinary service members with a tourist's understanding of the local culture. Yet it is the ordinary service members who have the most interaction with locals, particularly in a COIN scenario.

The problem with reliance on non-military personnel is well illustrated by the HTS program. Consider the main criticisms that have been levelled at it: it has been sharply criticised by the anthropological establishment for weaponising the discipline and, partially in consequence, has not proved able to recruit sufficient numbers of skilled scholars; it has resulted in deaths of (and killings by) HTT members allegedly because they were unprepared for a combat environment; it has also suffered from confused lines of reporting, oversight and authority (resulting in corruption and waste); and it has provided little help to the average service member on the front lines and so on. Relying on experts tasked primarily to advise commanders is wholly understandable in responding to a crisis situation. But in the longer term these programs need to be reimagined.

Our main conclusion can be summarised like this: the genuinely super soldier of the future will be defined not only by superior technology and improved physical capabilities, but also by a deeper understanding of the environment in which s/he functions. Some of countries fielding armed forces in Afghanistan are already exploring this model, including the Canadian, which put together teams analogous to HTTs but 'made up exclusive of uniformed military' along with government Foreign Affairs and Intelligence officers. Similarly Britain's Defense Cultural Specialist Unit is staffed predominantly by 'service military officers' (Lamb 2013, pp. 84–5).

In practical terms, of course, one cannot train every service member in every language of potential future war zones, but one could, for example, make strategic language acquisition and the fundamentals of social science important components of basic training, and make the developing of these skills and abilities a feature of daily life and a key consideration in promotion. Other possibilities include the following:

- Increase the priority on recruiting soldiers with social science backgrounds;
- Offer as a recruitment incentive an option to receive educational subsidies (including for the study of social sciences) before beginning a contracted term of service;
- Require all service members to complete CAOCL cultural training courses (which are already available online);
- Expand TRADOC and increase requirements for service members to complete programs as part of their training (with emphasis on strategic foreign languages and social science); and
- Institute a military reserve program aimed at professors capable of serving as strategic advisors.

The particular languages and regional expertise that service members develop could also be a factor in their assignment to units, so that particular units have collective strengths in performing in specific theatres. Finally, in nurturing these super soldiers, the military would be providing them with key skills that would facilitate their eventual return to civilian life. In this way the army may produce not just super soldiers but also better citizens. We admittedly do not venture deeply into the details of how this might be done – that must await future efforts. But we conclude that while this vision of the super soldier may be more prosaic than some of the alternatives on offer in this book, it is also one that is firmly grounded in the nature of current and likely future conflict scenarios, the changes in military thinking in recent years and the experiences (both positive and negative) of current programs. It is also one that potentially serves the individual and the polity as well as the military.

References

AAA Commission on the Engagement of Anthropology with the US Security and Intelligence Communities 2009, 'Final Report on the Armies Human Terrain System Proof of Concept Program', October 14, viewed on 3 October 2014, http://www.aaanet.org/cmtes/commissions/ceaussic/upload/ceaussic_hts_final_report.pdf.

AAA Executive Board 2007, 'Statement on the Human Terrain System Project', *American Anthropological Association*, October 31, 2007, viewed on 25 July 2014, http://www.aaanet.org/issues/policy-advocacy/Statement-on-HTS.cfm.

Abbe, A. and Gouge, M. 2012, 'Cultural Training for Military Personnel: Revisiting the Vietnam Era', *Military Review*, viewed on 3 June 2014, http://usacac.army.mil/CAC2/MilitaryReview/Archives/English/MilitaryReview_20120831_art005.pdf.

Bender, B. 2007, 'Efforts to aid US roil anthropology: Some object to project on Iraq, Afghanistan', *Boston.com* 8 October, viewed on 3 June 2014, http://www.boston.com/news/nation/washington/articles/2007/10/08/efforts_to_aid_us_roil_anthropology.

Bordin, J. 2011, 'A CRISIS OF TRUST AND CULTURAL INCOMPATIBILITY: A Red Team Study of Mutual Perceptions of Afghan National Security Force Personnel and U.S. Soldiers in Understanding and Mitigating the Phenomena of ANSF -Committed Fratricide-Murders', *N2KL Red Team Study*, viewed on 21 June 2014, http://www2.gwu.edu/~nsarchiv/NSAEBB/NSAEBB370/docs/Document%2011.pdf.

'Brochure for Understanding the Culture of Coalition Forces' (Excerpts from) 2012, *RadioFreeEurope/RadioLiberty* 13 September, viewed on 3 June 2014, http://www.rferl.org/content/excerpts-from-afghan-cultural-sensitivity-guide/24707518.html.

Brown, K. 2008, '"All They Understand Is Force": Debating Culture in Operation Iraqi Freedom', *American Anthropologist*, viewed on 3 June 2014, http://onlinelibrary.wiley.com/doi/10.1111/j.1548–1433.2008.00077.x/pdf.

Center for Army Lessons Learned (CALL) 2009, 'Escalation of Force: Afghanistan'. *CALL*, Ft. Leavenworth, KS, viewed on 21 July 2014, https://call2.army.mil/docs/doc5806/10-11.pdf.

Clinton, Y., Foran-Cain, V., McQuaid, J., Norman, C. and Sims, W. 2010, *Congressionally Directed Assessment of the Human Terrain System*, CAN Analysis & Solutions, November, viewed on 3 October 2014, https://info.publicintelligence.net/CNA-HTS.pdf.

Cocks, T. 2008, 'Are U.S. troops learning from cultural blunders in Iraq?', *Global News Journal RSS* 21 September, viewed on 3 June 2014, http://blogs.reuters.com/global/2008/09/21/are-us-troops-learning-from-cultural-blunders-in-iraq.

Cox, D. 2011, 'Human Terrain Systems and the Moral Prosecution of Warfare', *Parameters*, Autumn.

Dehghanpisheh, B. 2008, 'Scions of the Surge. *Newsweek*, March 2008, viewed on 25 July 2014, http://www.twcenter.net/forums/showthread.php?152883-Newsweek-Iraq-Fives-Years-Later-Scions-of-the-Surge-(great-read).

Department of the Army 2008, 'Field Manual 3-0: Operations', February, viewed on 14 August 2014, http://downloads.army.mil/fm3-0/FM3-0.pdf.

Fawcett, Major G. 2009, 'Cultural Understanding in Counterinsurgency: Analysis of the Human Terrain System'. Fort Leavenworth, Kansas: School of Advanced Military Studies.

Finney, Captain N. 2008, *The Human Terrain Team Handbook*, Fort Leveanworth, Kansas: Human Terrain System.

Grant, A. and Farrell, M. 2013, 'Approaches to Cross Cultural Competence for U.S. Marines in North Africa', *Social Science Research Network*, viewed 21 July 2014, http://papers.ssrn.com/sol3/papers.cfm?abstract_id=2251735.

Hamilton, Colonel S. 2011, 'HTS Director's Message', Military Intelligence Professional Bulletin, October-December, viewed on 25 July 2014, http://humanterrainsystem.army.mil/MIPB_Oct-Dec11.pdf.

Institute for Defense Analysis 2011, 'Human Terrain Team Study – Final Report', unpublished.

Jebb, C., Hummel, L. and Chacho, T. 2008, 'Human Terrain Team Trip Report: A "Team of Teams"', USMA Interdisciplinary Team in Iraq, viewed on 3 October 2014, http://www.usma.edu/chss/siteassets/sitepages/research/USMA%20HTT%20Study%20Report.pdf.

Keeley, L. 1996, *War Before Civilization*. New York: Oxford University Press.

Lamb, C., Orton, J., Davies, M. and Pikulsky, T. 2013, *Human Terrain Teams: An Organizational Innovation for Sociocultural Knowledge in Irregular Warfare*. Washington, DC: The Institute of World Politics Press.

Lewis, Lieutenant Colonel Brett G. "Developing Soldier Cultural Competency." U.S. Army War College Strategy Masters Research Paper, 2006. http://www.google.com/url?sa=t&rct=j&q=&esrc=s&frm=1&source=web&cd=1&ved=0CCYQFjAA&url=http%3A%2F%2Fwww.dtic.mil%2Fcgi-bin%2FGetTRDoc%3FAD%3Dada449393

&ei=XKKSVZ2gJ4-YyASn34CADg&usg=AFQjCNE7TK-L_cw_
zG4qzekHN3UqM1GcEg&sig2=-EvnQvEMVjc2Qp8O0AaHtw.

Long War Journal 2014, 'Green on Blue Attacks in Afghanistan: the Data', viewed on 29 July 2014, http://www.longwarjournal.org/archives/2012/08/green-on-blue_attack.php#data.

Matyszczyk, C., 'Obama: We're building Iron Man', viewed on 2 June 2014, http://www.cnet.com/news/obama-were-building-iron-man.

Memorandum 2010, 'Findings and Recommendations, AR 15–6 Investigation Concerning Human Terrain System (HTS) Project Inspector Complaints', TRADOC, 12 May 2010, viewed on 3 October 2014, https://app.box.com/s/2mv0g54xsr41aegwbw9i.

Nigh, N. 2011, 'CIWAG Study on Irregular Warfare and Armed Groups: An Operator's Guide to Human Terrain Teams', United States Naval War College, 2011.

Nordland, R. 2012, 'Culture Clash With Afghans on Display at Briefing', *The New York Times* 6 September, viewed 30 June 2014, http://www.nytimes.com/2012/09/07/world/asia/afghan-and-american-culture-clashes-at-center-stage.html.

Rahimi, S. and Rubin, A. 2012, 'Koran Burning in NATO Error Incites Afghans', *The New York Times* 21 February, viewed on 3 June 2014, http://www.nytimes.com/2012/02/22/world/asia/nato-commander-apologizes-for-koran-disposal-in-afghanistan.html?pagewanted=all&_r=0.

Rosenberg, M. 2012, 'Afghanistan's Soldiers Step Up Killings of Allied Forces', *The New York Times* 19 January, viewed 3 June 2014, http://www.nytimes.com/2012/01/20/world/asia/afghan-soldiers-step-up-killings-of-allied-forces.html.

Spitzer, K. 2012, 'Troops in Japan Told to Put a Cork in it', *Time* 28 November, viewed on 3 June 2014, http://nation.time.com/2012/11/28/troops-in-japan-told-to-put-a-cork-in-it.

Stanton, J. 2013, *US Army Human Terrain System*. Createspace Independent Publishing Platform (23 July).

Thompson, Captain. Interview with Barbara Gurgel. 10 May 2014.

Youssef, N. 2007, 'Pentagon: U.S. troops shot 429 Iraqi civilians at checkpoints', *McClathy DC* 11 July, viewed from 21 June 2014, http://www.mcclatchydc.com/2007/07/11/17836/pentagon-us-troops-shot-429-iraqi.html.

US Army Training and Doctrine Command 2006, 'Arab Cultural Awareness: 58 Factsheets'. *TRADOC*. Ft. Leavenworth, KS, 23 June 2014, http://fas.org/irp/agency/army/arabculture.pdf.

US Army Training and Doctrine Command 2008, *The US Army Study of the Human Dimension in the Future: 2015–2024*. *TRADOC*. Ft. Leavenworth, KS, viewed on 31 July 2014, http://www.tradoc.army.mil/tpubs/pams/p525-3-7-01.pdf.

Washington Post, 'US Troops in Afghanistan', viewed on 2 October 2014, http://www.washingtonpost.com/world/national-security/us-troops-in-

afghanistan/2014/05/27/41d8d60e-e5d1-11e3-afc6-a1dd9407abcf_graphic.
html.
Wunderle, W. 2007, 'Through the Lens of Cultural Awareness: A Primer for US
Armed Forces Deploying to Arab and Middle Eastern Countries', *Combat
Studies Institute Press*, Ft. Leavenworth, KS, viewed on 3 June 2014, http://
usacac.army.mil/cac2/cgsc/carl/download/csipubs/wunderle.pdf.

Chapter 5

Morally Enhanced Soldiers: Beyond Military Necessity

Ryan Tonkens

In the first instance, any discussion of biogenetically enhanced soldiers should be a powerful wake-up call. That we might *need to* change anything substantial about human beings so that they can be (more) effective in their militarist roles is deeply unsettling. It implies that war is becoming so complicated, rapid, and foggy that human soldiers in their unaltered state cannot adequately keep up with its pace and demands. Perhaps this has always been true, although the situation seems exacerbated in the context of contemporary warfare, accompanied by ever-evolving technological capabilities.

Military necessity demands that, in order for militaries and individual soldiers to keep up – in order for them to have a legitimate chance at protecting themselves and the citizens they represent, and overcoming their adversaries – they need to design technologies *and themselves* to be better than their opponents. In contexts where opponents are less advanced in this regard, this imperative will be less pronounced. Yet, while the continued development of military technologies is a centuries-old endeavour, and it has long been recognised that military forces with 'the strongest' and 'the best' soldiers and technologies are more likely to be victorious, new developments in robotics, nanotechnology and biotechnology mean that these technologies are becoming more sophisticated, rapidly evolving and intrusive.[1]

Even in the face of heated debate on the ethics of human enhancement (in both military and civilian contexts), some such enhancements are already available, and, if things keep going as they are, more are likely to emerge in the future. Even if we assume that there is nothing *inherently* morally unacceptable about (safe) human enhancements, the morality of specific modes of enhancement depends on the context in which they are used, and the end(s) that they are being used to achieve. For example, while there may be nothing inherently morally unacceptable about enhancing humans to be more patient through pharmaceutical means (say), enhancing individuals to be more patient so that they could be more effective

---

1   We might speculate that eventually there will be no significant role for humans to play in warfare, at least not anywhere near the front lines, in which case warfare might become futile, absurd (Krishnan 2009).

torturers would seem to be morally problematic. This is because torture itself is morally unacceptable, and deliberate contributions to promoting human torture are by extension morally unacceptable as well.[2] Enhancing the patience of snipers *may* be less problematic, and enhancing the patience of soldiers so that they are less likely to make hasty decisions when engaging with enemy combatants (and thus less likely to fail to discriminate between legitimate combatants and innocent civilians) is likely to be less unacceptable still. Or, 'enhancing' soldiers to exhibit a lack of empathy for the plight of other human beings (even if only in their enemies) would promote a lack of discrimination, potentially loosen concerns for proportionality and promote immoral behaviour. This is true even if having less empathy for one's opponent (whether one is a sniper, torturer or whatever) could promote military proficiency.

While warfare in general is ugly, an enterprise that routinely sparks the most despicable of human behaviour, it is also taken to be necessary, especially given that threats of attack and violence from the outside are pervasive. The ethical issues that emerge at the intersection of human enhancement and warfare are tricky to grapple with in part because of the blurriness of the moral acceptability of the ends of warfare taken more generally: on one hand, it is largely assumed that *defending* oneself (either as an individual or as a nation) is morally acceptable, and thus that (at least) *some* military action is permissible, or at worse a necessary evil in an imperfect world, where our wellbeing and way of life is constantly vulnerable to outside threat. On the other hand, it is also largely agreed that killing others is wrong (for reasons other than defence), and that the colossal harms that come about as a result of warfare are undesirable, and thus that war is something that, in the long term, we ought to work to mitigate and eliminate altogether. (To say that war is 'inevitable' is to misuse language, however unlikely it seems that putting an end to war once and for all is an achievable goal, given the current state of the world.) As long as military actions conform close enough to the former justification, they are taken to be morally justifiable. Something has gone awry, however, if we lose sight of the latter goal *altogether*, that is, if we ignore the idea that war should not occur at all.

An analogy may help to clarify this important point: *if* it turns out that there is a moral imperative to be vegetarian, then we should become vegetarians. This would imply the need to design a context where vegetarianism is promoted and encouraged, where the long-term goal is to curtail and eliminate the consumption of meat. *If* it were unrealistic (or harmful, or imprudent) to give up eating meat once and for all at the present time, then we might allow the continued consumption of meat (as a necessary evil), while *also* demanding that we simultaneously work

---

2 I recognise that the ethics of human torture remains contentious, but appeal here to the fact that human torture is recognised in international laws of war as being unacceptable. There are also some recent philosophical treatments of the ethics of torture that offer a compelling case against the permissibility of human torture, on both consequentialist and deontological grounds (for example Bufacchi and Arrigo 2006).

towards remedying this barrier to moral behaviour, in the long term. There might be good reasons, say, for treating the nonhuman animals that we intend to eat better than we currently do, even if widespread and sustained vegetarianism is unlikely to become a reality *today*. In keeping in line with the moral imperative to be vegetarian, we could (*should*) at the same time be taking measures to promote vegetarianism, and avoid as much as possible promoting behaviour (or policies, or the like) inimical to vegetarianism. From within a carnivorous culture, the need to reorient that culture so as to be a vegetarian one is founded on the (*ex hypothesi*) moral imperative towards vegetarianism. Even if we are carnivores right now because of the (less than ideal) state of ourselves and the world in which we live, losing sight of this long-term goal would nevertheless be unacceptable.

A central element of my argument in this chapter is that soldier 'enhancements' that are inimical to or inconsistent with the long-term goal of peace (and peaceful conflict resolution) are morally problematic, and that enhancements that *only* contribute to militarist ends are morally suspect: exclusive appeal to military necessity is not sufficient grounds for establishing the moral acceptability of specific military enhancements, since we should also be working towards establishing a world where war is not 'necessary', and such enhancements are problematic to the extent that they are inconsistent with that goal.[3] Specific enhancements that do not meet this condition are to that extent morally problematic, even if they are in line with military proficiency.

The Imperative of Enhancing the Morality of Warfighters

In a series of recent works, Persson and Savulescu (2008, 2012a, 2012b, 2013) have been calling for research into and the development of drugs and biotechnologies for the purpose of morally enhancing humans. They believe the future of our species depends on doing so, specifically in the face of rapid developments in cognitive enhancement. For them, the threat of what they term Ultimate Harm (UH), that is, activity that, in the extreme, ends worthwhile life on Earth, is real and ubiquitous in a world where people will (*do*) have the enhanced cognitive capacities to know how to bring it about, and the resources available to bring it about if they so desire.

The urgent goal is to make sure that everyone that has the capacity to bring about Ultimate Harm does not desire to do so; or at least that their moral motivation is powerful enough to prevent them from using their advanced cognitive capacities in deplorable ways. *One* way to promote this goal, for these authors, is to initiate sustained research and development into drugs that will make humans morally better (just like we have developed drugs to make humans better at remembering things, staying awake longer, feeling better about themselves and so on), drugs that

---

3   This should not be mistaken for the stronger claim that soldier enhancements that are justified by (exclusive) appeal to military necessity or military proficiency *are necessarily* morally unacceptable.

could (for example) lower aggression, truncate discrimination, curb xenophobia and so on. The advent of widespread cognitive enhancement (in tandem with our growing information culture and technological capabilities) will make it possible for an increasing number of people to be 'clever' enough to be able to plan the demise of the human species. As this pool of people grows, it is bound to then include more people who would also *desire to* initiate UH, and have the resources available for doing so successfully.

With or without cognitive enhancement, warfare and human behaviour therein sets the stage for many potential avenues for initiating UH (and other less extreme yet nevertheless catastrophic harms). Although Persson and Savulescu do not discuss the role of moral enhancement in the military context specifically, one of the more obvious ways that Ultimate Harm *could* come about would be in the context of war, where the resources necessary for initiating UH are sometimes (or even often) available, and in a context that has been the arbiter of the greatest atrocities that the history of the human species has been witness to. It is not (just) the 'typical' antisocial, ill-intentioned, cognitively enhanced individual that will attempt to end the world or the species, but rather (also) individuals that are in certain respects *much better-placed to do so* (for example those that have access to advanced military technology and that have the training to use such technology effectively).

Fending off the perils of cognitive enhancement requires moral enhancement (Persson and Savulescu 2008). Part of the point being made here is that this is especially true for those that participate in the theatre of war, as this is a context that is already riddled with immoral and inhumane behaviour, and this is a context from which UH is more likely to emerge, compared to (most) civilian contexts. Morally enhancing those that have the resources and capacity to initiate UH might be crucial for preventing UH before it occurs. If we *just* cognitively enhance warfighters, and not also enhance their morality, then the risk of some such warfighters misusing their military resources and expertise increases as well (to some, presently indeterminate, extent). Or, if we enhance soldiers without paying any attention to the potential impacts of such enhancements, and without counteracting their potential effects with moral enhancements, then the risks highlighted by Persson and Savulescu become more likely. If this is right, an urgent context for widespread moral enhancement is the military context, especially those militaries that have access to weapons of mass destruction, and especially those that intend on enhancing their soldiers to be more militarily efficient.[4]

---

4   Arkin (2009 and 2010) has argued for the development of (semi-)autonomous lethal robotic systems to replace or supplement human soldiers on the battlefield, mostly since they could presumably behave less immorally and less inhumanely than (many) human soldiers characteristically do. Arkin presents a convincing case, I think, to show that human soldiers are, as a group, in much need of moral improvement. Given that human soldiers are susceptible to behaving immorally in the theatre of war, *one* way to attempt to improve the moral calibre of war would be to make it so that soldiers are not (as) susceptible in this way. And, *one* way to do this may be to replace human soldiers with sufficiently

Missing Condition for Morally Acceptable Soldier Enhancements

Given that the dominant militarist mindset focuses (often exclusively) on military necessity and proficiency, it is not surprising that the considerations emphasised above have yet to emerge in the debate surrounding soldier enhancement. This militarist orientation is obvious in recent policy recommendations, like the ones offered by Mehlman, Lin and Abney (2013) for example. While much of their account is compelling and will likely be helpful for progressing the debate in this area, there are a notable omission in their account, which render their recommendations at best incomplete.

In order for a soldier enhancement to be morally acceptable, according to these authors, it needs to satisfy the following nine conditions: (1) it needs to serve a 'legitimate' military purpose; (2) use of that enhancement (in this way) must be 'reasonably necessary' in order to attain the legitimate military end set out in (1); (3) the benefits of this enhancement must be greater than the costs (to the soldiers, and non-combatants as well (Mehlman, Lin and Abbey, p.122)); (4) the enhancement must not compromise or usurp the soldier's dignity; (5) such enhancements should pose minimal burdens to the soldier (for example through being temporary or readily reversible); (6) the soldier consents to the enhancement; (7) the reasons motivating the pursuit of this enhancement should be made transparent to the public, to the extent possible – presumably the authors mean to say transparency to the extent that the military ends for which the enhancement is being used are not thereby compromised; (8) the risks associated with the enhancement are not to be too concentrated, but rather spread out as widely as possible (that is, we are to avoid focusing the burden only on a few); and (9) those that prescribe the enhancements (among others, perhaps) are to be held accountable for misuse.[5]

---

autonomous lethal robotic systems. I have been critical of Arkin's overall position elsewhere (Tonkens 2012), although I do not rehash those points of contention here. What I do want to highlight, however, is the idea that Arkin jumps from the moral imperfection of human soldiers to the creation of autonomous lethal robotic systems (as their replacements and/ or moral police) without pausing to ask whether there may be other ways to make human warfighters less immoral, rather than creating autonomous lethal robotic systems to replace them. Indeed, one reason why we might want to pause here first is that the development and use of autonomous lethal robotic systems has attached to it very many sticky moral and practical issues that makes their development and use morally suspect, as things currently stand. Given that we already do allow humans to fight in wars, if there were safe and effective ways to make these people more moral, then we may not need to dive into the messy territory of lethal robotic systems in the first place (at least not for *this* reason). There is enough here to warrant serious consideration of such alternatives. One such alternative would be morally enhancing soldiers through pharmaceutical means.

5   To this list we would do well to add the condition that military enhancements should not significantly lessen (or eliminate) the soldier's autonomy or capacity to be held responsible for their actions (Wolfendale 2008).

Here I will omit discussion of most of Mehlman et al.'s conditions, but would like to mount a criticism motivated by the incompleteness and misleading nature of the first and second of their conditions listed above (taken in tandem), both of which have to do with what makes justification for any given soldier enhancement legitimate.

The major problem with these two conditions is that they assume an *exclusive* militarist perspective at their foundation. One implication of this is that they end up becoming simultaneously too broad and too narrow: adhering to these conditions means that any military enhancement that does not have a legitimate *military* purpose is automatically ruled out, and enhancements that are in line with military necessity but inconsistent with (for example) a long term pacifist agenda are ruled in (assuming they meet the other conditions as well). Asking for military enhancements to conform to military necessity and proficiency ignores the fact that *not being in line with military necessity is not necessarily problematic, even within the military context.* Given that war and behaviour in war is morally acceptable insofar as it is grounded in some way in self-defence or the protection of the vulnerable, the prevention of significant harm to oneself or one's nation from undue outside threat, and *this* only becomes a legitimate justification for warfare once we recognise that warfare *should* not occur at all, then we are obliged to look *also* to non-militarist ends when determining which behaviour ought to be allowed in this context.

It seems reasonable to require that all military enhancements serve legitimate military ends in the sense that they need not be superfluous or futile in this regard. What I want to suggest here is that, even if the rest of the conditions noted above are satisfied, this would not be sufficient for establishing the moral permissibility of specific military enhancements. What is missing here is acknowledgement that war *should* not occur, and that war is only tolerated (that is, accepted as morally permissible) because war still does occur. Just because there is good reason to enhance our militaries in the name of proficiency and military necessity, it does not follow that doing so while *ignoring* the inherent undesirability and harmfulness of human warfare is justifiable or acceptable.

For example, enhancing the *aggressiveness* of soldiers is orthogonal to their sound moral character, taken objectively to include both behaviour within and outside of the military context, and with the long-term goal of truncating and eliminating the occurrence of war (and peaceful conflict resolution). This is true even if enhancing aggressiveness in this way might, in some cases, be consistent with military proficiency. If humans were less aggressive, were less prone to attempt to resolve their differences by resorting to aggressive behaviour, then the occurrence of war would be less pervasive. Because enhancing for elevated aggression is inimical to the (long-term) goal of peaceful conflict resolution, even if 'military necessity' supports such an enhancement initiative, doing so is morally problematic. Drugs that aim to enhance humans only in ways that promote military proficiency *and* are inconsistent with non-militarist agendas are morally problematic because we should be working towards establishing the preconditions for a warless world (and the people that are being 'enhanced' are not *just* soldiers, and may at some point re-enter into civilian society). Where military proficiency and sound moral character diverge, the argument I have presented here would suggest that military proficiency, pursued *in such ways*, ought to be forfeited. This is reinforced also from the direction of attempting to prevent the occurrence of Ultimate Harm.

Admittedly, there is some sense in which the *moral* enhancement of military personnel is oxymoronic: we would not want to go too far in the direction of morality, lest such military professionals spontaneously throw down their arms, or become disinterested in deliberately targeting other humans with lethal force, which would undermine military proficiency to some extent. (This may sound ideal to some people, but is not a *realistic, short-term* goal.) Yet, if there were ways to enhance a sense of moral responsibility, or enhance psychological or physiological aspects of human soldiers that make them less likely to commit wartime atrocities, or the like, then these ought to be pursued, alongside other soldier enhancements.

Many of the points being made in this chapter are not new ones – although they have yet to appear in the literature on military enhancement. The imperative of morally enhancing soldiers has long been recognised, and is an important professional aim (Wertheimer 2013). Many countries now have as part of their standard military training *at least some* training in ethics (that is, Canada and the United States). Yet, this pedagogical approach may not be sufficient *on its own*, especially for combating the potential perils of cognitive enhancement and the increased complexity of modern warfare. Morally enhancing soldiers through (for example) pharmaceutical means, would help to combat the threat of Ultimate Harm, and have the subsidiary benefit of making war less inhumane and immoral.

Conclusion

My account here is motivated by two central claims. The first is that Ultimate Harm is a real threat, and this threat more forcefully comes from military personnel or those that have military backgrounds (and access to the requisite information, equipment and so on). The second point is that moral legitimacy of warfare is grounded in the laudable aims of protecting oneself and the vulnerable from undue outside threat – war is something that is tolerated, given the current imperfect state of the world.

By focusing exclusively on military necessity, we lose sight of the idea – shared by warists and pacifists alike – that war *should* not occur at all. When designing and using drugs aimed at enhancing soldiers, we pay homage to this goal by making sure that all such enhancements are not inimical to it, and, where soldier enhancement is widespread, to counteract potential negative effects by also morally enhancing those same soldiers. To the extent that we pursue the development of military enhancements, we should also be pursuing the development of military *moral* enhancements, whereby we take seriously the fact that war in general is only acceptable *if* war exists, but also that war ought not to exist.

One potential objection to the argument presented here is that this proposal is too restrictive, such that all or just about all military enhancements would be ruled out, which would go in the face of (dire) military necessity. However, my argument is not that each pharmaceutical soldier enhancement must, by itself, be both militarily enhancing and morally enhancing if it is to pass the test. The crucial point is rather that enhancements ought not to be inimical to important non-military ends, that is, the long-term goal of mitigating and eliminating war. If military enhancements compromised or undid the positive effects of corresponding military *moral enhancements* (whether they are pedagogical, pharmaceutical or otherwise), *these* enhancements would not satisfy the minimal threshold of not being inimical to a long-term, realistic pacifist agenda, and would thereby be presumptively morally wrong. If these enhancements made warfare more immoral and less humane than it already is, we would be moving in the opposite direction from which we should be moving. One of the assets of the view being defended here is that it pays due recognition to *both* long-term pacifist goals *and* the demands of military necessity in the world in which we currently live.

A related potential objection is that the account articulated here is naïve. Even if we *should* limit our military enhancements to those that are not inimical to a long-term pacifist agenda, we do so at our own peril, given that our enemies will not be so scrupulous. The first thing to note in response is that this line of objection would have us accept *anything* insofar as not accepting it would put us in jeopardy or make us vulnerable. *This* would surely be too lax. The second thing to note is that the philosophical debate on the ethics of military enhancement presupposes that ethics is something that should be at play in this context, at least to some extent. If it turns out that ethics does not matter *at all* in the context of war, or in the face of unscrupulous enemies, then volumes like the one in which this chapter appears would become redundant. The third thing to note is that many military enhancements, especially those that satisfy Mehlman et al.'s list of conditions, *will be morally innocuous*, that is, will not be inimical to or inconsistent with the pursuit of a long-term pacifist agenda. Enhanced aggression would be an exception here. Enhanced patience (discussed in the introduction) would likely pass *for reasons other than to breed enhanced torturers*. Enhancements for things such as memory capacity, wakefulness or alertness, intelligence and physical stamina would likely satisfy the condition set out in this chapter.

Worth emphasising is the idea that we should not just be concerned that some of *us* might be in a position to initiate UH, but, perhaps more pressingly, that some of our enemies might desire to do so as well. And, part of the drive towards military enhancement (of our own military personnel) is precisely to combat this sort of external threat, as much as possible. Given that avoiding UH is a pressing concern, one might argue that *mandatory* moral enhancement of everyone, *including all soldiers, on all sides*, is required. And yet, assuming that our enemies would be unlikely to morally enhance themselves just in case we ask them to, perhaps the best we can do at this point is to continue to attempt to protect ourselves from such potential harms (through military means, where necessary), *and* make sustained and substantial efforts towards creating a world where such action is no longer necessary. One aspect of this would be to make sure that none of *our* soldier enhancements are blatantly inimical to this long term goal.

References

Arkin, Ronald C. 2010, 'The case for ethical autonomy in unmanned systems'. *Journal of Military Ethics* 9(4): 332–41.

Arkin, Ronald C. 2009, *Governing Lethal Behavior in Autonomous Robots*. Boca Raton, FL: CRC Press.

Bufacchi, Vittorio and Jean Arrigo. 2006, 'Torture, terrorism and the state: A Refutatilon of the ticking-bomb argument'. *Journal of Applied Philosophy* 23.3: 355–73.

Mehlman, Maxwell, Patrick Lin and Keith Abney. 2013, 'Enhanced warfighters: A policy framework', in *Military Medical Ethics for the 21st Century*, eds Michael L. Gross and Don Carrick. Farnham: Ashgate, 113–26.

Krishnan, Armin. 2009, *Killer Robots: Legality and Ethicality of Autonomous Weapons*. Aldershot: Ashgate.

Savulescu, Julian and Ingmar Persson. 2012, 'Moral enhancement, freedom and the god machine'. *The Monist* 95(3): 399.

Persson, Ingmar and Julian Savulescu. 2013, 'Getting moral enhancement right: the desirability of moral bioenhancement'. *Bioethics* 27(3): 124–31.

Persson, Ingmar and Julian Savulescu. 2012, *Unfit for the Future: The Need for Moral Enhancement*. Oxford: Oxford University Press.

Persson, Ingmar and Julian Savulescu. 2008, 'The perils of cognitive enhancement and the urgent imperative to enhance the moral character of humanity'. *Journal of Applied Philosophy* 25(3): 162–77.

Tonkens, Ryan. 2012, 'The case against robotic warfare: A response to Arkin'. *Journal of Military Ethics* 11(2): 149–68.

Wertheimer, Roger. 2013, 'The morality of military ethics', in *Empowering Our Military Conscience: Transforming Just War Theory and Military Moral Education*, ed. Roger Wertheimer. Farnham: Ashgate, 159–96.

Wolfendale, Jessica. 2008, 'Performance-enhancing technologies and moral responsibility in the military'. *The American Journal of Bioethics* 8(2): 28–38.

PART II
General Problems and Consequences

Chapter 6

Enhanced Warfighters as
Private Military Contractors

Armin Krishnan

Human enhancement is coming. Several high-level policy papers in the US and Europe indicate that it will soon be technologically feasible to enhance human performance in key areas such as cognitive enhancement, sensory enhancement, physical enhancement and genetic enhancement. Of course, one of the driving forces in human enhancement technology is military research aimed at creating 'super soldiers', who are smarter, stronger and more resilient than ordinary human soldiers. But the ethical and legal implications resulting from 'super soldiers', and human enhancement in society more broadly, are daunting. For numerous reasons it could be very difficult for the modern armed forces of Western democracies to transition from a traditional to an 'enhanced' military. A likely solution for dealing with this dilemma is to deploy permanently enhanced warfighters covertly in small numbers. The idea is that each enhanced soldier would be an 'army of one' – a weapons system with unprecedented lethality and resilience. The first enhanced warfighters would thus be a select group of Special Operations Forces (SOF) and other clandestine personnel (for example paramilitary intelligence personnel). However, since these elite forces are in many ways closely connected to the top segment of the private security sector, comprised of a few elitist private military companies (PMCs), and since a lot of the human enhancement technology can be expected to quickly end up on black markets and in society in general, elite private security contractors are likely to be the second group of warfighters to receive controversial permanent human enhancements. Through covert use and the further privatisation of military force, Western democracies could potentially dodge the serious ethical dilemmas while also taking advantage of 'super soldiers' earlier than they could otherwise. This chapter will discuss the types of controversial enhancements that could be in reach for PMCs, indicate some of the ethical and legal issues regarding these enhancements, outline the present role of PMCs in warfare and sketch the possible long-term implications of PMCs fielding enhanced warfighters. It will be concluded that human enhancement is likely to further increase the role of PMCs and other substate actors in warfare, which would be accompanied by a relative decline of national armed forces and nation-states that would signify the ascent of a neomedieval world.

Controversial Human Enhancements

Human enhancement technology is currently at R&D stage and much of it will be expensive and limited to technologically advanced armed forces. But certain types of human enhancement technologies could be still in reach or be attractive to private military contractors. As PMCs lack the resources to spend billions of dollars in R&D and since they could not easily acquire cutting-edge military technology, they could only utilise human enhancements by hiring former enhanced military personnel or if it was available on commercial or black markets at affordable prices. Private security contractors could not hope to afford 'Iron Man'-type power suits with jet packs or guided missiles built into them, nor would they be allowed as civilians to operate such weapons systems. However, there are some forms of human enhancement that could be cheap and available enough within ten years, especially when it comes to pharmacological enhancements, genetic enhancements and brain–machine interfaces.

Pharmacological Enhancements

The use of drugs by soldiers and military contractors during combat operations has already become commonplace and it has already raised some red flags with respect to the ability of drugged troops to exercise good judgement. There are essentially two types of drugs used by the military and contractors: medical prescription drugs, some of which could have a performance-enhancing effect in terms of keeping soldiers awake and alert for longer periods of time (for example modafinil, Dexedrine), or of reducing fear and psychological trauma (SSRI), and illegal drugs like cocaine or synthetic drugs that can have similar effects, but that are more addictive and have even worse effects on the health of users. About 6 per cent of US active duty troops are taking anti-depressants (Murphy 2012) and overall drug usage could be even more widespread in the private sector. There have been several public cases where private military contractors in Iraq and Afghanistan were caught abusing alcohol, illegal drugs, as well as pharmaceuticals. In a court case against Blackwater a witness testified in a court in Virginia that 'he purchased steroids, human growth hormones, and testosterone for Blackwater employees and his observation of rampant drug use among Blackwater employees. Initially, Blackwater paid for the steroids from company funds' (Isenberg 2010). The general problem is that there is no consistent drug testing with respect to private contractors. Indeed, the US State Department has become so concerned about the problem that they now require semi-annual drug testing of their 1,600 career employees and contractors in Afghanistan and Israel, following an audit that had indicated 'deficiencies' in previous drug-testing practices (Taylor 2013). However, more drug testing would not fix the problem of psychotropic prescription drugs. Since one in five Americans take at least one psychiatric medication (Friedman 2013) and since PTSD is also rampant amongst security contractors, it can be assumed that many of them would be on some sort of psychotropic drugs when deployed. The psychiatric drugs used to battle depression or PTSD can result in 'mania', 'aggression' and 'violent behaviour', as indicated in the side effects disclaimers.

Genetic Modification

Some of DARPA's human performance enhancement programs look into the genetic modification of soldiers. Its 'Living Foundries' program is meant to reduce the costs of gene therapy and to create 'tools for rapid physical construction of biological systems, editing and manipulation of genetic designs' (DARPA 2012). In other words, DARPA is researching the technology for changing the genetic make-up of soldiers through gene therapy to make them stronger, give them more endurance and to make them more resilient to biological and environmental threats or conditions. The main idea of gene therapy is to introduce new DNA into the body that fixes the damaged DNA through a specifically engineered recombinant virus that changes the DNA contained in all body cells. Although still expensive, the technology exists and several diseases have been successfully treated with gene therapy, including immune deficiencies, hereditary blindness, haemophilia, blood disease, fat metabolism disorder, cancer and Parkinson's disease. Gene therapy could be potentially used for genetic improvement. It is quite possible that some athletes may have already engaged in 'gene doping' in the recent 2012 Olympics by injecting themselves with genes that promote muscle growth or otherwise enhance their performance (Naish 2012). 'Tomorrow's soldiers could be able to run at Olympic speeds and will be able to go for days without food or sleep, if new research into gene manipulation is successful', according to an article in the *Daily Mail* (Gayle 2012). From a longer term perspective, even more radical genetic modifications such as the creation of human–animal hybrids with unique capabilities is possible in principle. A report by the European Union *Global Governance 2025* discusses the genetic modification of humans and argues that the 'direct modification of DNA at fertilisation is currently widely researched with the objective of removing defective genes; however, discussions of future capabilities open the possibility for designing humans with unique physical, emotional or cognitive abilities' (2010, p. 56). The Oxford transhumanist Nick Bostrom has pointed out the theoretical possibilities of equipping humans with new features: '[t]he current human sensory modalities are not the only possible ones … Some animals have sonar, magnetic orientation, or sensors for electricity and vibration; many have a much keener sense of smell, sharper eyesight, etc. … There is no fundamental block to adding say a capacity to see infrared radiation or to perceive radio signals and perhaps to add some kind of telepathic sense by augmenting our brains with suitably interfaced radio transmitters' (Bostrom 2005, p. 7). Without doubt it would give warfighters a tremendous advantage on the battlefield if they had a sensory perception that goes far beyond the enemy's capabilities in this regard. However, radical genetic modification of humans is far beyond existing science, would not be possible in adults and would thus require embryonic intervention, thereby creating a 'pre-specified warrior class' (Ford and Glymour 2014, p. 46).

Brain Implants

The US military is interested in implants for brain stimulation and as brain–machine interfaces (BMIs). Brain stimulation can enhance the mental performance of soldiers such as allowing them to remain focused for longer periods of time, but it could also be used for suppressing feelings, cravings and thoughts. Up to now more than 50,000 people have received deep brain stimulation (DBS) implants for therapeutic reasons, such as depression, Gilles de la Tourette syndrome, alcoholism and Alzheimer's disease (Blank 2013, pp. 32–3). The US military has recently invested $70 million into developing a DBS implant for curing PTSD. 'The project builds on expanding knowledge about how the brain works; the development of microelectronic systems that can fit in the body; and substantial evidence that thoughts and actions can be altered with well-placed electrical impulses to the brain' (Regalado 2014). The potential for behavioural control through brain stimulation has been thoroughly demonstrated in animal studies and is achieved through 'simple stimuli that either simulate known physical (sensory) signals for specific actions, or provide a "rewarding" impulse' (JASON 2008, p. 5). The next step is to develop a two-way communication interface with the brain. Neuroscience is working hard in decoding the human brain in order to develop a seamless brain–machine interface (BMI) to enhance sensory and motor abilities. Again, scientific research in this area has been ongoing for decades and a BMI is in reach within years. There are three obvious military applications of BMIs: 1) cognitive augmentation, which alarms soldiers to threats they do not consciously perceive and to respond before a conscious intention is formed, which could cut down effective response times by as much as 7–8 seconds (Kasanoff 2012); 2) synthetic telepathy, which would allow soldiers to communicate silently amongst each other or with some distant command post (Piore 2011); and 3) thought-controlled weapons (for example fighter jets) that take advantage of the greater cognitive abilities of the human brain compared to computer vision while significantly reducing human response times to threats (Aym 2012). Non-invasive EEGs make it already possible to monitor some brain activity and to do 'brain finger printing' (determine whether the brain recognises a particular stimulus). However, the kind of mind-reading capability necessary for thought-controlled weapons and synthetic telepathy would probably require a brain implant that can more precisely decode the electrical signals in the brain. This advanced technology is not only being developed by DARPA, but also by private corporations like Google and Intel. Scientists from Intel believe that PCs will be controlled by thought with a BMI by 2020 (Gaudin 2009). Google is envisioning a BMI to give its customers the ability to do Internet searches by thought and is working on a brain chip to allow disabled people to steer their wheelchairs by thought (Sherwin 2013). According to the *Wall Street Journal*, '[e]ventually neural implants will make the transition from being used exclusively for severe problems such as paralysis, blindness or amnesia ... When the technology has advanced enough, implants will graduate from being strictly repair-oriented to enhancing the performance of healthy or "normal" people. They will be used to improve memory, mental focus (Ritalin without the side effects), perception and mood (bye, bye Prozac)' (Marcus 2014). When the technology becomes available to the general public it would be also available to security contractors, who would have a strong incentive to get enhanced in that way.

Problems in Transitioning to a Post-human Military

Many of DARPA's ambitious human enhancement programs, especially those that would permanently modify human soldiers, have been widely criticised for their potential ethical and legal implications. For example, DARPA's genetic and neuroscience projects have raised the public concern that the agency could be secretly building 'Frankenstein armies' with remotely controlled brains, a concern that was acknowledged and also dismissed by former DARPA director Anthony Tether in an interview (Shachtman 2007). But while the technology is rapidly leaping ahead, ethical and legal concerns regarding certain types of enhancement (for example pharmacological and genetic enhancement and brain implants) still exist and they could make it difficult for traditionally minded and constitutionally limited armed forces to transition to a post-human military.

Human Experimentation

The need for human experimentation necessary for developing certain human enhancement technologies is indeed a major impediment to the development and deployment of the technology, as indicated by Tether (Shachtman 2007). All human experiments have to pass through ethics review boards and this substantially slows down the R&D process. There is also a requirement to get 'informed consent' from test subjects (with a few legal exceptions), which legally prevents the possibility of generally forcing experimental enhancements on soldiers. Although the Greenwall report notes: 'the instinct for self-preservation is likely to lead warfighters to grasp at any means of improving their chances of surviving battle, including exposing themselves to the risks of experimentation in order to gain access to experimental enhancements' (Lin et al. 2013, p. 49), there have been many cases where soldiers have refused to be medicated with experimental drugs. One major example is of course the US Army's use of an untested anthrax vaccine before the Gulf War of 1991, which may have been at least partially responsible for the 'Gulf War syndrome' that affected over 250,000 US Gulf War veterans. Gary Matsumoto has made a convincing case that an unlicensed adjuvant (squalene) in the anthrax vaccine was chiefly responsible for severely damaging the health of many soldiers, who had been ordered to be vaccinated against anthrax and who had not given any informed consent, a normal requirement for experimental treatment (Matsumoto 2004). Forced anthrax vaccination created so much controversy that it led to a class-action lawsuit by members of the military who had refused vaccination and who were subsequently demoted, dismissed or court-martialled, which they ultimately won and which changed the policy toward voluntary anthrax vaccination in the military (Elbe 2010, p. 89). Brookings' Peter Singer acknowledged that 'the Pentagon's real-world record with things like Agent Orange, the Tuskegee Syphilis tests, above ground testing of atomic bombs and nerve agents, and Gulf War Syndrome certainly don't inspire the greatest confidence that everything will turn out perfectly', also quoting a SOF soldier: '[b]eing a guinea pig doesn't settle well with me' (Singer 2008). In particular, 'genetic intervention would require morally intolerable experimentation' (Ford and Glymour 2014, p. 46) and could be outlawed altogether. Since human enhancement technologies are fairly new, there is little knowledge available regarding any long-term health risks to individuals. It could thus take decades before certain enhancements can be declared safe and considered no longer experimental in nature.

Ethical Concerns with Respect to Genetic Engineering and
Brain–Machine Interfaces

A report by the National Academy of Sciences suggests that soldiers should be selected for duty based on their genetic make-up (NAS 2009, p. 20). Although this can result in a more capable military, it is also easy to see the ethical issues that would arise from such a practice of genetic discrimination. The genetic selection of soldiers goes completely against the Western democratic military tradition that is based on equal opportunities and personal achievement. Furthermore, 77 per cent of OECD countries have already adopted policies that prohibit genetic screening for non-medical purposes (Hayes 2008, p. 5). However, the alternative of bringing individuals on the same level genetically through genetic modification of healthy individuals could be ethically even more problematic. Military effectiveness would dictate that it cannot be left to individual soldiers to decide whether or not to accept body modifications as unenhanced soldiers could not operate well alongside enhanced ones. This problem could necessitate coercing soldiers to accept enhancements, which would obviously infringe upon the civil liberties of soldiers, a concern also raised by the Royal Society's report on military neuroscience applications (Royal Society 2012, p. 27). Some pharmacological enhancements, brain-implants for brain stimulation and brain–machine interfaces raise also some very troubling questions with respect to freedom of will, capability for moral responsibility and human dignity. Michael Tennison and Jonathan Moreno have argued that '[i]f a warfighter is allowed no autonomous freedom to accept or decline an enhancement intervention, and the intervention in question is as invasive as remote brain control, then the ethical implications are immense' (Tennison and Moreno 2012, p. 2). Democratic societies may not be very comfortable with the idea of soldiers getting equipped with BMIs that could allow military commanders to remotely manipulate their brains.

International Law Limitations for an Enhanced Military

Some researchers have already pointed out that there could be serious legal obstacles to developing or fielding enhanced warfighters. For example, Patrick Lin has argued that any modification that would turn soldiers into uncontrollable weapons, for example a 'berserker-drug', would have to be considered a violation of the Geneva Conventions (2013). The creation of human–animal hybrid soldiers could be 'repugnant to the conscience of mankind' and thus violate the Martens clause. He also suggests that 'super soldiers' could be subject to other sources of international law such as the Biological and Toxin Weapons Convention, which makes it illegal to even develop 'biological agents' as weapons of war, which is not necessarily limited to microbial agents. Brain–machine interfaces also create new legal dilemmas, which have been explored by Stephen White. Although White notes: '[t]he development and use of weapons coupled with brain–machine interfaces most likely does not violate international criminal law', he also points out that they raise 'novel issues in the jurisprudence of war crimes' (White 2008, p. 178). A major problem would be to establish the guilt of an individual who has killed civilians with a weapon that was controlled by a brain–machine interface. Could the individual in question be made responsible on the basis of their thoughts, which goes against Western legal tradition? To what extent can thoughts be considered voluntary acts or treated as actions? Similarly, as is the case with autonomous weapons, brain–machine interfaces could be declared illegal if they break the chain of command responsibility and if they therefore make it impossible to hold anybody accountable for war crimes. As a result, Western militaries may have to forego the possible advantages of certain forms of human enhancement and may therefore lose their competitive technological edge in warfare. More likely is that governments may dodge the various ethical and legal issues through privatisation, as they previously did with respect to several other ethically problematic national security activities like secret human experimentation, torture and assassination. The following sections will sketch the current role of PMCs in contemporary warfare and will argue that they are likely to embrace controversial human enhancement technologies, possibly before Western militaries could do so officially.

PMCs and Warfare in the Early Twenty-First Century

It has been frequently pointed out that PMCs have become so integral to US military, diplomatic and intelligence operations that the US government no longer seems able to manage without them. There is now a $218-billion-dollar-a-year global security services industry with an annual growth rate of 7.4 per cent (Freedonia 2011). The industry has already transformed the way the West intervenes in developing countries and it has also created serious challenges for the state control of military force.

Some Characteristics of PMCs

PMCs have a corporate structure, utilise contemporary management philosophies and operate as profit-driven businesses: some are subsidiaries of publicly traded companies (for example MPRI is a subsidiary of L3, Dyncorp was acquired by CSC), some are standalone, privately owned companies (for example Academi or Triple Canopy). They tend to be relatively small with only a small number of permanent staff and a large number of former military professionals, who can be called up when the company gets a major contract. PMCs often hire foreign nationals or subcontract foreign companies to augment their limited manpower. Since they need to be able to operate globally, they need to have the necessary communications and logistics capabilities, which is, in the age of the Internet and satellite phones, no longer a significant barrier. Apart from that, PMCs do not have much overhead: there is no need for a big corporate headquarters or for maintaining weapons and ammunitions stockpiles. They can typically operate on a shoestring budget as they do not use much heavy weaponry or other heavy equipment. Even armoured vehicles and helicopters that have been used by Blackwater in Iraq are rather uncommon for the industry because of the logistical burden and the increased cost that cuts into the profits of PMCs. The main capital of any PMCs is the military skill and expertise of their employees, who are rented out to government and corporate clients.

Advice and Training

PMCs tend to offer a range of services. The most common type of service is advice and training, which in some cases even extends to participation in combat. This industry has been traditionally dominated by UK and US companies (Kinsey 2006, p. 1). The first modern PMC that was organised in a corporate structure and offered advice and training to militaries in Latin America, the Middle East, Africa and South East Asia was Watchguard International, which was founded by former SAS officer David Stirling. Keeni Meeni Services (KMS), Saladin Security, Defence Systems Ltd, and Executive Outcomes (EO) followed in its footsteps (Dunigan 2011, p. 2). PMCs have intervened in a number of conflicts, often tilting the balance of power in favour of their clients, most notably in former Yugoslavia, Angola and Sierra Leone. British PMCs have also provided advice to and collected intelligence on behalf of numerous multinational corporations, mostly in the extractive sector (mining and oil companies). Typical services, as provided by a leading British company in the field, Control Risks Group, are investigations, political risk consulting, hostage negotiation, counterterrorism training and other security-related training.

Military Security Services

With the beginning War on Terror and the subsequent occupations of Afghanistan and Iraq US and UK, PMCs have rebranded themselves as 'private security companies'. They focus on securing objects and personnel in conflict zones, of mostly, but not exclusively, government clients. In 2003 tens of thousands of foreign contractors were flowing into Iraq to do everything from basic logistics to providing security for VIPs such as Ambassador and American 'Vice-Roy' Paul Bremer. Companies like Blackwater, Dyncorp and Triple Canopy defined the PMC business in the first decade of the twenty-first century. In 2007 the US Department of Defense estimated that there were 180,000 contractors in Iraq, outnumbering the 160,000 US troops deployed there. Estimates of the number of security contractors ranged as high as 30,000 (Dunigan 2011, p. 1). Although it is very hard to find any solid data since even the US military had problems keeping track of them, it seems that most of them were not US citizens, but came from all over the world to participate in the so-called 'Baghdad boom'. Peter Singer notes that they came from at least 30 different countries (Singer 2007). Not surprisingly, there were many newcomer companies that lacked the experience and the qualified personnel, but were hired anyway because of the high demand for security in Iraq. In addition, the business has been transformed by the myriad of local firms that sprung up to work as subcontractors for Western PMCs. These local companies are in reality often little more than warlords, militias and thugs for hire. During this decade of private security growth a select group of companies and individual contractors found themselves moving away from mundane security services tasks into the much more shadowy realm of secret intelligence operations and covert action.

PMCs in Special Operations and Intelligence

The George W. Bush administration was very open-minded when it came to outsourcing core military and intelligence functions. They even considered hiring Blackwater to hunt and kill al Qaeda members worldwide as the CIA lacked such a capability at the time (Boot 2013, p. 346). Although nothing came out of this unusual initiative, Blackwater personnel (under the new name Xe Services) was operating on behalf of the CIA in Pakistan, handling drones at the Pakistani air base Shamsi and reportedly collecting intelligence in preparation of US drone strikes (Scahill 2009). Using a PMC for this work was dictated by the problem that Pakistan would not allow the US to have US military personnel operate within their country. The *New York Times* suggested 'that their involvement in the operations became so routine that the lines supposedly dividing the Central Intelligence Agency, the military and Blackwater became blurred' (Risen and Mazetti 2009). US security contractors are also heavily involved in the War on Drugs with the Pentagon recently awarding $3 billion in contracts to private security firms to assist with global counternarcotics and counterterrorism operations (Ackerman

2011). Apart from Blackwater, there is a select number of American PMCs, most notably Dyncorp, Triple Canopy and GK Sierra, that work very closely with the CIA, Joint Special Operations Command and the State Department on missions that require deniability or a low profile, which is often the main reason why they are hired in the first place. Non-disclosure clauses are routinely written into PMC contracts, prohibiting the security contractors to talk to the media (Kelly 2011). In addition, some security contractors are given special security clearances that allow them to participate in highly classified operations, also known as Special Access Programs or 'black ops' (Scahill 2009).

Implications of Human Enhancement and PMCs

Western PMCs are already very well positioned for playing a key role in twenty-first-century warfare because so much of it has become 'cloak-and-dagger'. PMCs not only offer greater flexibility, additional capability and sometimes reduced cost, but first and foremost they offer 'plausible deniability'. PMCs can intervene in situations where it would be difficult for the regular military to do so openly. They can keep a low profile because they are technically civilians and thus do not have to wear uniforms or national insignia. Their presence in a conflict zone or neutral country can remain secret and this can help avoiding public discussions or controversies that would arise from official troop deployments. In the future the use of PMCs instead of the official military could make it easier for governments to use enhanced warfighters without having to directly confront all the ethical dilemmas regarding human enhancement in the military.

Lacking Accountability

It has been pointed out that '[t]he relative impunity of private military companies (PMCs) is increasingly puzzling' (Leander 2011, p. 467). Even some of the private security sector's greatest advocates complain about the general lack of regulation of the industry and the lacking accountability of military contractors. Most problematic about private security contractors is that they are in legal terms neither mere civilians nor full combatants, which means in practice that they can claim exceptional status with respect to domestic laws while not being fully covered by international law. Heather Carney argues: 'PMFs [private military firms] are now performing many functions historically undertaken by national militaries. Yet, not only do their employees lack the protections and benefits enjoyed by military members, they are also not constrained by the same obligations, such as the U.S. Military Code and the Geneva Conventions' (Carney 2006, p. 319). Private security contractors have gotten away with gross human rights violations in the past, some of which had resulted in highly publicised scandals, for example Dyncorp employees engaging in human trafficking in the Balkans, CACI employees engaging in torture at Abu Ghraib and Blackwater employees shooting

17 civilians in Nisour Square, Baghdad. On occasion, the US government has provided legal immunity to its security contractors and has used its diplomatic leverage to prevent their prosecution overseas, as in the case of Raymond Davis who was arrested in Pakistan for murder in 2011. It seems that governments have so little enthusiasm for regulating contractors chiefly because they are using them for controversial purposes that are better not publicly discussed or too closely examined in court rooms. PMCs may also not be bound by current or future US domestic laws against genetic modification or forced implantation, as many of them are headquartered overseas or generally employ foreigners. Important international law such as the Biological and Toxin Weapons Convention that may constrain regular militaries from developing and fielding enhanced warfighters may not even apply to them. Private security contractors could therefore take advantage of certain human enhancements even if the regular military could not.

'Frankenstein Armies' Through the Private Sector Backdoor

The actual use of human enhancement technologies in warfare could be driven more by private actors rather than regular militaries. Certainly, the technology would be first available to a few SOF soldiers, who carry out clandestine missions that are not officially acknowledged, which would allow Western governments to go secretly beyond existing ethical and legal boundaries. However, human enhancement would quickly leak into the private sector. Enhanced SOF soldiers could be recruited by PMCs or some private security contractors would privately choose to get enhanced with technologies available commercially or on black markets. The incentives for doing so would be very strong. Even if certain human enhancement methods such as genetic modification would be strictly regulated in the US or other Western countries, there would be surely a supplier in another country where there was less regulation or capability for enforcing laws against human genetic engineering. Security contractors that have modifications that give them unique capabilities would command the highest salaries and their employment could give PMCs a huge competitive edge over their opponents (and business competitors). It can be expected that PMCs and other more nefarious private actors will be pushing the envelope in terms of human enhancement. A Congressional hearing led by Representative Brad Sherman in the House Foreign Affairs subcommittee in June 2008 discussed 'the diplomatic and security implications of the spread of 'genetics and other human-modification technologies'. Several expert witnesses addressed biotech's potential for creating 'super soldiers, super intelligence, and super animals', debating the possibility that biotech 'could put agents of unprecedented lethal force in the hands of both state and non-state actors' (Hayes 2008, p. 1). The greatest concern one can have about the proliferation of military human enhancement technology is the possible creation of privately run 'Frankenstein armies' that operate worldwide with little or no accountability. Even under the best scenario of governments outlawing all human genetic engineering and non-medical human implantation with BMIs, there remains a great possibility that the technology will be illegally developed and used, and that in the more distant future some humans will be modified in such radical ways that they may no longer even be considered human or be granted human rights.

Neomedievalism and Human Enhancement

The proliferation of human enhancement technology to private actors would fit very well into a larger trend towards neomedievalism in international relations. Since controversial human enhancement could be adopted by private actors faster than by national militaries, it would hasten their ongoing relative decline. Similarly to warfare in the late Middle Ages and the chartered trading companies of the eighteenth century, privately controlled force could once again assume a dominant role in warfare. Eugene Smith claims that two self-reinforcing trends eventually put private war entrepreneurs out of business by the end of the nineteenth century: '[t]he growth of bureaucratically mature states capable of organising violence created increasingly strong competition for private military corporations. At the same time, states began to recognise that their inability to control the actions of these private organisations challenged state sovereignty and legitimacy. The result was that the utility of the private military corporation as a tool of state warfare disappeared ... until recently' (Smith 2002, pp. 107–8). With conventional interstate war getting out of fashion and with a new focus on messy low-tech conflicts in developing countries, PMCs seem like a better option than the expensive and risky deployment of national armies. PMCs enable Western countries to intervene unofficially with little political commitment. Small teams of highly trained mercenaries can already offer to their clients SOF-like capabilities to achieve strategic objectives. In the future it could be possible to make every single warfighter several times more effective and lethal than today, which means that manpower and possession of major weapons systems is no longer an entry barrier for competing in war. If such capabilities were offered on the private market, even small states and substate actors, including city states, multinational corporations and even rich individuals, could then wage war against each other, creating more chaos and instability. Analyst Parag Khanna claims: '[t]his diffuse, fractured world will be run more by cities and city states than by countries ... Add in sovereign wealth funds and private military contractors and you have the agile geopolitical actors of a neomedieval world' (Khanna 2009, p. 91). In other words, the coming decades could usher in some form of neomedievalism, which would undermine or undo much of the existing world order and create a political landscape that is infinitely more fragmented and complex.

Conclusion

The technology for substantially enhancing human performance and to modify humans will be available within 10 years. Western militaries are interested in fielding enhanced warfighters, but it will be difficult for them to overcome ethical and legal concerns. In order to avoid these issues while taking advantage of human enhancement, they may opt to deploy enhanced warfighters covertly or to simply hire enhanced individuals as private military contractors. As a result, national or international efforts aimed at preventing human genetic engineering or the use of non-medical brain implants could be undermined. Furthermore, human enhancement could make PMCs much more competitive compared to national militaries and this may result in them assuming a dominant role in twenty-first-century warfare, comparable to the condottieri of the late Middle Ages. It could be therefore difficult, if not impossible, to stop the eventual creation of post-human troops or 'Frankenstein armies'. This would create more dangers for civilians in conflict zones and could undermine the individual freedom, moral responsibility and human dignity of the enhanced warfighters.

References

Ackerman, S. 2011, 'Pentagon's war on drugs goes mercenary', *Wired Magazine*, 22 November, viewed 30 June 2015, http://www.wired.com/2011/11/drug-war-mercenary.

Aym, T. 2012, 'Thought-controlled weapons', *World Issues 360*, 9 February, viewed 27 May 2014, http://www.worldissues360.com/index.php/thought-controlled-weapons-4293.

Blank, R.H. 2013, *Intervention in the Brain: Politics, Policy, and Ethics*. Cambridge, MA: The MIT Press.

Boot, M. 2013, 'Afterward: The CIA and Erik Prince', in Prince, E., *Civilian Warriors: The Inside Story of Blackwater and the Unsung Heroes of the War on Terror*. New York: Penguin.

Bostrom, N. 2005, 'Transhumanist values', *Review of Contemporary Philosophy*, 4.

Carney, H. 2006, 'Protecting the lawless: human rights abuses and private military firms', *The George Washington Law Review*, 74: 317–44.

DARPA 2012, 'Living Foundries', *Biological Technologies Office*, viewed 23 May 2014, http://www.darpa.mil/Our_Work/BTO/Programs/Living_Foundries.aspx.

Dunigan, M. 2011, *Victory for Hire: Private Security Companies' Impact on Military Effectiveness*. Stanford, CA: Stanford University Press.

Elbe, S, (2010), *Security and Global Health*. Cambridge: Polity.

Ford, K. and Glymour, C. 2014, 'The enhanced warfighter', *The Bulletin of Atomic Scientists*, 70(1): 43–53.

Freedonia 2011, 'Global security service to exceed $218 billion in 2014, says Freedonia', *asmag.com*, 17 March, viewed 12 June 2014, http://www.asmag.com/showpost/11443.aspx.

Friedman, R.A. 2013, 'A dry pipeline for psychiatric drugs', *The New York Times*, 19 August, viewed 11 June 2014, http://www.nytimes.com/2013/08/20/health/a-dry-pipeline-for-psychiatric-drugs.html.

Gaudin, S. 2009, 'Intel: Chips in brains will control computers by 2020', ComputerWorld, 19 November, viewed 30 June 2015, http://www.computerworld.com/article/2521888/app-development/intel--chips-in-brains-will-control-computers-by-2020.html.

Gayle, D. 2012, 'Army of the future: soldiers will be able to run at Olympic speed and won't need food or sleep with gene technology', *Mail Online*, 12 August, viewed 21 May 2014, http://www.dailymail.co.uk/sciencetech/article-2187276/U-S-Army-Soldiers-able-run-Olympic-speed-wont-need-food-sleep-gene-technology.html.

Hayes, R. 2008, 'Is there an emerging consensus on the proper uses on human genetic technologies?', Testimony of Richard Hayes, House Foreign Affairs Committee Subcommittee on Terrorism, Nonproliferation and Trade, 19 June.

Isenberg, D. 2010, 'PSCs on Drugs', *The Huffington Post*, 23 September, viewed 12 June 2014, http://www.huffingtonpost.com/david-isenberg/pscs-on-drugs_b_736590.html.

JASON 2008, Human Performance, The MITRE Corporation, McLean, VA, March, viewed 30 June 2015, available at: https://fas.org/irp/agency/dod/jason/human.pdf.

Kasanoff, L. 2012, 'DARPA's new "brain–computer interface" makes you a pattern recognition machine', *DigitalTrends.com*, 1 October, viewed 10 June 2014, http://www.digitaltrends.com/cool-tech/this-is-your-brain-on-silicon/#!W4X6D.

Kelly, S. 2011, 'Confessions of a private security contractor', *CNN.com*, 27 December, viewed 11 June 2014, http://security.blogs.cnn.com/2011/12/27/confessions-of-a-private-security-contractor.

Khanna, P. 2009, 'Neomedievalism', *Foreign Policy*, 172: 91.

Kinsey, C. 2006, *Corporate Soldiers and International Security: The Rise of Private Military Companies*. London: Routledge.

Leander, A. (2010), 'The paradoxical impunity of private military companies: authority and the limits to legal accountability', *Security Dialogue*, 41(5), 467–90.

Lin, P. 2013, 'Could human enhancement turn soldiers into weapons that violate international law?', *The Atlantic*, January, viewed 23 May 2014, http://www.theatlantic.com/technology/archive/2013/01/could-human-enhancement-turn-soldiers-into-weapons-that-violate-international-law-yes/266732.

Lin, P., Mehlman, M.J. and Abney, K. 2013, *Enhanced Warfighters: Risks, Ethics, and Policy*. San Luis Obispo, CA: California Polytechnic State University.

Marcus, G. 2014, 'The future of brain implants: how soon can we expect to see brain implants for perfect memory, enhanced vision, hypernormal focus or an expert golf swing?', *The Wall Street Journal*, 14 March, viewed 12 June

2014, http://online.wsj.com/news/articles/SB1000142405270230491490457943559298178052.

Matsumoto, G. 2004, *Vaccine A: The Covert Government Experiment that's Killing our Soldiers and Why GI's are Only the First Victims*. New York: Basic Books.

Murphy, K. 2012, 'A fog of drugs and war', *The Los Angeles Times*, 7 April, viewed 28 May 2014, http://articles.latimes.com/2012/apr/07/nation/la-na-army-medication-20120408.

Naish, J. 2012, 'Genetically modified athletes: Forget drugs: there are even suggestions some Chinese athletes' genes are altered to make them stronger', *The Daily Mail*, 31 July, viewed 30 June 2015.

National Research Council 2009, *Opportunities in Neuroscience for Future Army Applications*. Washington, DC: The National Academies Press.

Piore, A. 2011, 'The Army's bold plan to turn soldiers into telepaths', *Discover Magazine*, 20 July, viewed 27 May 2014, http://discovermagazine.com/2011/apr/15-armys-bold-plan-turn-soldiers-into-telepaths.

Regalado, A. 2014, 'Military funds brain–computer interfaces to control feelings', *MIT Technology Review*, 29 May, viewed 11 June 2014, http://www.technologyreview.com/news/527561/military-funds-brain-computer-interfaces-to-control-feelings.

Risen, J. and Mazzetti, M. 2009, 'Blackwater guards tied to secret C.I.A. raids', *The New York Times*, 10 December, viewed 30 June 2015, http://www.nytimes.com/2009/12/11/us/politics/11blackwater.html?pagewanted=all&_r=0.

Roco, M.C. and Bainbridge, W.S. (eds) 2002, *Converging Technologies for Improving Human Performance: Nanotechnology, Biotechnology, Information Technology, and CognitiveScience*. Arlington, VA: NSF/DOC sponsored report.

Royal Society 2012, *Brain Waves Module 3: Neuroscience, Conflict and Society*, Royal Society, viewed 27 May 2014, https://royalsociety.org/policy/projects/brain-waves/conflict-security.

Scahill, J. 2009, 'The secret US war in Pakistan', *The Nation*, 7 December, viewed 30 June 2015, http://www.thenation.com/article/secret-us-war-pakistan#.

Shachtman, N. 2007, 'Be more than you can be', *Wired Magazine*, Issue 15.03, viewed 22 May 2014, http://archive.wired.com/wired/archive/15.03/bemore.html.

Sherwin, A. 2013, 'Google's future: microphone's in the ceiling and microchips in your head', *The Independent*, 9 December, viewed 26 May 2014, http://www.independent.co.uk/life-style/gadgets-and-tech/features/googles-future-microphones-in-the-ceiling-and-microchips-in-your-head-8993990.html.

Singer, P.W. 2007, 'Can't win with 'em, can't go to war without 'em', Brookings Institution Policy Paper 4, http://www.brookings.edu/~/media/research/files/papers/2007/9/27militarycontractors/0927militarycontractors.pdf.

Singer, P.W. 2008, 'How you can be all you can be: A look at the Pentagon's five step plan to make *Iron Man* real', Brookings Institution, viewed 26 May 2014, http://www.brookings.edu/research/articles/2008/05/02-iron-man-singer.

Smith, E.B. 2002. 'The new condottieri and US policy: The privatization of conflict and its implications for security', *Parameters*, 32(4): 104–19.

Taylor, G. 2013, 'Drug tests ordered for State Department contractors in Israel, Afghanistan', *The Washington Times*, 26 May, viewed 10 June 2014, http://www.washingtontimes.com/news/2013/may/26/drug-tests-ordered-for-state-department-overseas-c.

Tennison, M.N. and Moreno, J.D. 2012, 'Neuroscience, ethics, and national security: The state of the art', *PLoS Biology*, 10(3): 1–4.

White, S. 2008, 'Brave new world: Neurowarfare and the limits of international humanitarian law', *Cornell International Law Journal*, 41: 177–210.

Chapter 7

Super Soldiers and Technological Asymmetry

Robert Simpson

Any technological innovation that confers a welcome benefit upon a positionally advantaged military force carries a corresponding cost for positionally disadvantaged military forces – at least temporarily – while the relevant technologies are distributed in a decidedly uneven fashion. And it is one of the ethicist's jobs to worry about these costs, which are borne by the 'have-nots' in global conflict. For those who espouse an ethics of Absolute Pacifism, there will not be much to say about the costs associated with emerging military technologies, (since – for Absolute Pacifists – no war can be rendered morally justifiable thanks to a novel military technology's involvement in it). For all others interested in the ethics of warfare, however, the ethical significance of the costs accompanying new military technologies have to be examined case-by-case.

What should we say, then, about the potential costs and possible downsides of soldier enhancement technology? For the technologically disadvantaged military force (henceforth, the *Underdog*), one of the costs that accompanies the development of effective soldier enhancement technologies by an opposing, technologically advanced military force (henceforth, the *Superpower*) is entirely generic: the Superpower's large advantages over the Underdog – in weaponry, communications, transport and so on – are bolstered by a further type of advantage, courtesy of which the Superpower becomes better able to cement its stranglehold over the coercive use of violent force in the global political arena. And the upshot of *this* is an entrenchment of the circumstances of asymmetric political violence, that is, the kind of dynamics which drive Underdogs towards the use of terrorist violence against civilian populations. The significance of technological asymmetry as a precursor to such violence has been elucidated by others, and I will not recapitulate that discussion here (see Killmister 2008; Fabre 2012, pp. 239–82). Instead I will attempt to map out a distinctive ethical problem which is generated by soldier enhancement technologies, albeit one that relates to a more general family of ethical issues in asymmetric warfare.

Several authors have argued that when major technological disparities separate the opposing sides in a political conflict, these disparities render the use of lethal violence *by* Superpowers *against* Underdogs unjustifiable. I will explain this view in the first section. If one accepts this view, however, it remains unclear what uses of violent force *are* justifiable for Superpowers in such conflicts. And,

as I will argue in the second section, the suggestion of people like Paul Kahn –
that in these sorts of conflicts, Superpowers ought to eschew *warfare* in favour
of *policing* – is unconvincing. Why? Here is the worry in brief. In approaching
such conflicts as occasions for *policing* wrongdoing, rather than engaging in full-
scale combat, the Superpower's individual personnel relinquish the relatively
unthreatened position that they would otherwise occupy in a combat scenario with
the Underdog. And thus, the shift from combat to a policing approach cannot be
obligatory for the Superpower, since the very basis of the rationale which is meant
to make the Superpower's shift to a policing approach obligatory is the fact that
the Underdog's forces *do not* pose any threat to the Superpower's forces. Or so I
will argue. The link with the topic of this collection will then become clear in the
final section. My claim will be that the advent of effective soldier enhancement
technology transforms the circumstances of threat and risk that obtain in conflicts
between Superpowers and Underdogs, in a way that may enable the Superpower
to undertake a policing approach in a political conflict with Underdog, but *without*
their personnel relinquishing the relatively unthreatened position they would
otherwise enjoy in full-scale combat with Underdog forces. And if that is right,
then the advent of effective soldier enhancement technology supports the view
that I outline in the first section – about the obligation of Superpowers to eschew
combat in favour of policing – by removing the objection that I present in the
second section.

   In a sense, then, what I will be arguing is that the development of effective
soldier enhancement technologies can generate ethically significant costs on both
sides of the military technology divides. The cost for the Underdog is to be faced
with even greater technological disadvantages, which make the possibility of
effective uses of violent force in political conflict even more remote. The cost for
the Superpower – assuming they purport to abide by reasonable ethical constraints
on armed conflict – resides in the fact that the advantages gained via soldier
enhancement technology also generate onerous responsibilities in violent political
conflict. What sort of responsibilities? In short, those that come with taking on the
duties of policing.

Technological Asymmetry and the Paradox of Riskless War

Several authors have recently defended something like the following claim: where
there are large disparities in the combat capabilities of parties involved in an armed
conflict, these disparities greatly shrink the range of circumstances under which it
is morally justifiable for the advantaged party to carry out lethal attacks against the
disadvantaged party (for example, see Dunlap 1999; Kahn 2002; Galliott 2012a,
2012b, 2015; Steinhoff 2013; Simpson and Sparrow 2014). The idea, put simply,
is that the permissibility of killing in war depends upon there being a 'fair fight',
in the sense that the belligerent opponents cannot be grossly unevenly matched in
their warfighting capabilities. If this view is correct, it casts a further shadow of

doubt across the moral justifiability of a military superpower like the United States carrying out lethal attacks on enemy combatants in many of the conflicts that it has been involved in over the last 15 years, such as in Yemen, Afghanistan, the trans-Saharan region and in the horn of Africa.[1] Even if we assume that the standardly acknowledged ethical constraints on conduct in war are painstakingly honoured in such conflicts – even if it were true in these conflicts that US forces were limiting the damage that they were inflicting to what was necessary for the achievement of their legitimate military aims, *and* only targeting enemy combatants *and* taking significant further measures to minimise harm to non-combatants – it might still be the case that US forces were not justified in carrying out lethal attacks against opposing combatants in these conflicts.[2] Why is that? In short, because the disparities between the combat capabilities of the US and their opponents here are far too great; the circumstances of mutual endangerment between the opposing fighters – which is a necessary condition for the justifiability of killing in war – simply do not obtain. Or so the argument goes.

In view of these considerations, Paul Kahn (2002) has argued that states with highly well-equipped military forces face a 'paradox of riskless warfare'. Such states will naturally aim to achieve superiorities in combat capability which reduce, as far as possible, the risks incurred by their personnel in combat situations. But to the extent that such aims are realised – for example where a Superpower like the US succeeds in greatly mitigating the risk to their personnel in combat situations – the Superpower *ipso facto* delegitimises its employment of lethal force against the technologically disadvantaged Underdog, in view of the 'fair fight' constraint noted above. Kahn's view has obvious ethical implications for the Superpower using unoccupied weaponised vehicles to conduct lethal attacks on opposing Underdog forces. The drone operator can kill enemy combatants from afar, while incurring no reciprocal risk. And this is, of course, the key consideration that makes drone warfare a strategically appealing combat option.[3] For a proponent of

---

1   Of course the point that I am making here doesn't apply to the United States alone; I mention these asymmetric conflicts involving the US just because the US is, by all measures, the most powerful and technologically advanced military force in the world today.

2   Here I am gesturing toward the two core principles of *jus in bello* that figure in all standard accounts of just war theory – namely, (i) Proportionality (roughly: damage inflicted must be limited to what is necessary for the achievement of legitimate military ends), and (ii) Discrimination (roughly: belligerents must distinguish between combatants and noncombatants and target only the former) – along with the supplementary principle endorsed (for example) by Michael Walzer (2006, pp. 151–9), that combatants must take measures aimed at minimising accidental harm to noncombatants, even if doing so carries significant costs with regards to their own safety.

3   I should note that some authors, like Bradley Strawser (2010), have argued that there is – over and above the manifest strategic advantages that come with the use of weaponised drones – in fact a moral *duty* for states to employ drones, in order to reduce risks to their own personnel. Note also that some other authors, like Beauchamp and Savulescu (2013), have defended a similar conclusion, ostensibly on the grounds that states will be more ready

Kahn's view, however, the lack of reciprocal risk in the use of weaponised drones is precisely the thing which renders lethal drone attacks ethically unjustifiable.

Jai Galliott has recently defended a similar conclusion, via a somewhat different route. His focus is on how the technological disparities between a drone-equipped Superpower and its Underdog opponents can preclude any adequate ethical justification for the decision to resort to armed combat in the first place. On all standard accounts of *jus ad bellum*, the waging of war – even war waged in the pursuit of an uncontroversially just cause – must be treated as a last resort. A state cannot justifiably initiate combat, then, unless it has previously exhausted the other strategic avenues that may be pursued in order to achieve whatever legitimate aims might (putatively) justify its use of coercive political violence. Galliott's point is that in cases where Superpowers face Underdogs, it will seldom (if ever) be the case that we can credibly characterise full-scale combat as a last resort for the Superpower. Because of the Superpower's enormous technological advantages, it will typically be at least *possible* for it to redress the aggression to which it is responding, without resorting to full-scale combat. So even if we assume that other constraints on justice in the resort to war are honoured, Galliott says, the use of drone warfare is (often) impermissible, since it results from the more powerful state's failing to treat war as a last resort (Galliott 2012a, pp. 62–4).

If the conclusion is correct – if large disparities in military technology ethically preclude lethal combat – then what uses of violent force *are* justifiable for a Superpower, A, in responding the aggression of an Underdog, B? According to Kahn, in such cases A should eschew *warfare* in favour of *policing*. Suppose B has carried out acts of military aggression towards A, or that it has committed humanitarian atrocities that are so egregious as to justify A's armed intervention. Under a warfare paradigm, A's aim will be to forcefully overwhelm B's capacity to use violent force in turn – for example by destroying B's military hardware and/or personnel – to the point where B's aggression can be decisively repelled, or its humanitarian atrocities prevented. By contrast, if A approaches the situation via a policing paradigm, its aims will be at least somewhat narrower, for example to apprehend culpable wrongdoers among B's political or military leadership, and to carry out some process aimed at holding them formally accountable for their wrongdoing.

The invocation of a policing paradigm already assumes that the advantaged party has a decisive upper-hand in its capacity to exert violent force, such that there is no need for it to win this advantage over the wrongdoers. The aim of policing – where would-be wrongdoers are *already* decisively outmatched in their capacity to exert violent force – is to subdue, apprehend and try the renegade actors who choose to engage in violent wrongdoing nevertheless. Note two further important differences between the paradigms. First: in war, violent force may

---

and willing to carry out ethically meritorious wars of armed humanitarian intervention in cases where the use of drones can mitigate or eliminate risks to the personnel of the intervening state.

be directed against *all* enemy combatants. In policing, by contrast, violent force must be directed toward only those who have (or who are reasonably suspected to have) violated a prohibition whose violation itself supplies a justifying basis for the use of violent force. Second: in war, uses of lethal force are permitted outside circumstances of imminent self-defence, for example pilots can bomb an opponent's military outpost, killing enemy combatants who do not pose an imminent threat to anyone's life. In policing, by contrast, the use of lethal force is restricted to circumstances of defending against an imminent threat – the police officer may only fire upon a suspected wrongdoer if s/he believes (reasonably) that s/he is presently endangering another's life. Granted, there are all sorts of ways in which a description of these differences could be qualified or more painstakingly formulated. The point is that where in warfare there is a general license to use lethal force against some specified class of persons – namely, enemy combatants – in policing there is no such general license. The prerogatives involved in policing relate to the use of sub-lethal violence for the purposes of law-enforcement. As in normal social intercourse, the justifiable use of *lethal* violence in policing is limited to circumstances of defence against an imminent threat to someone's life.[4]

I should acknowledge that views like Kahn's and Galliott's, about the ethical ramifications of technological asymmetry in war, are founded upon the contested assumption that opposing fighters are *in general* morally justified in using lethal force against one another in war. One may wonder, then, what the upshot would be if we followed those like Jeff McMahan (2009), who say that even in war the justifiable use of lethal violence is – as in normal social intercourse – limited to circumstances of self-defence against an imminent threat. My view is that a version of Kahn's paradox of riskless warfare arises even under McMahan's more restrictive ethical framework for thinking about killing in war. The kind of case in which McMahan accepts the justifiability of killing in war is the case where the soldier doing the killing is engaged in self-defence against an imminent lethal threat, posed by an aggressor fighting in the name of an unjust cause. In

---

4 In saying that there is a general license to use lethal force against enemy combatants in war, I do not mean to suggest that in war there is a *completely unqualified* license to kill enemy combatants. At a minimum, the *jus in bello* principle of proportionality imposes firm limits upon this license. As Tony Coady says, the 'entitlement to injure and kill [enemy combatants] is restricted by its necessity for furthering the war aims that are legitimated by your just cause, and when attacks upon them are no longer required by those aims, then the normal respect for human life should resume and be exhibited in your conduct' (Coady 2008, p. 157). My point is merely that there is a difference between war and policing, in that only in the former is the general respect for human life (provisionally) suspended, with regards to a specified class of persons (that is, enemy combatants). Though there are certain things that the police officer is in general permitted to do, which the rest of us are not in general permitted to do, 'killing enemies' is not one of them. The permissibility (and/or excusability) conditions of killing people are much the same for the police officer as for the rest of us, differing in practice primarily by virtue of the fact that police officers face lethal threats much more often than the rest of us.

this sort of case, as much as in any other, asymmetries in military technology can reduce the degree to which the Underdog's soldier poses a lethal threat to the Superpower's soldier. And so – even for the soldier fighting in self-defence against unjust aggression – there is a point where this disparity is so large, and (correspondingly) where the opponent's threat is so negligible, that the use of lethal force against the aggressor loses whatever moral justifiability it may otherwise have possessed. In McMahan's framework the initial range of cases of permissible killing in war is narrower than in a conventional just war theoretic framework. But, plausibly, there remains a common structure – across the two frameworks – to the way killings in war can be rendered impermissible due to technological asymmetries.

The Perils of Policing

Although I have considerable sympathy for Kahn's account, at the same time something about it seems to me implausible. Where soldier S1 is fighting for a technologically advanced military Superpower, A, and soldier S2 is fighting for a much less well-equipped Underdog, B, it is not clear why it should follow that the relational paradigm structuring the engagement between S1 and S2 is *police-versus-criminal*. Even if it becomes unfitting to regard the engagement between S1 and S2 as one of *combatant-versus-combatant*, this does not yet entail that a policing paradigm adequately describes the ethical contours of the interaction between S1 and S2.

Suppose, for example, that the Superpower, A, is the aggressor in the conflict, for example suppose A is using military force in pursuit of an unjust aim, like territorial occupation, whereas the Underdog, B, is using military force in an attempt to counter A's aggression. The problem, in that case, is that Kahn's approach still assigns S1 the role of 'police' and S2 the role of 'criminal'. To say that A would be justified only in policing B's conduct, rather than engaging in full-scale combat against B, is to overlook the most important ethical fact in the neighbourhood, namely, the fact that A's personnel are not justified in exerting violent force against B's personnel *in any form*. If large technological asymmetries alter what forms of violence are justifiable for the Superpower, in anything like the way Kahn suggests, it may yet turn out that this only obtains where the Superpower has some kind of (defeasible) justification for exerting violent force in the first place.

But even in that case, something in Kahn's view seems awry. Suppose that Underdog B is the aggressor against Superpower A, and that A is clearly justified in principle (under standard tenets of just war theory) in using violent force to repel the aggression. And suppose also that B's armed forces are amassed in a military encampment near A's borders (or, say, embassy), but not yet in the process of launching an attack against A. In such a case, can A's forces take the initiative and launch a lethal attack on B's encampment? If A is restricting itself to *policing* B's wrongdoing, as opposed to engaging in full-scale combat with B, then the answer is surely 'no'. After all, if members of a police squadron know that the building across the street is occupied solely by people who are planning to carry out a killing spree, they – the police – cannot justifiably respond to this by bombing the building to smithereens and thus killing all inside. They can try to apprehend the killers in advance, or else, once the violent acts are initiated, they can use lethal force to stop them. But pre-emptive lethal strikes have no place in any morally defensible form of police work. Returning to the military context, then: if policing really *is* the appropriate framework for understanding the moral character of A's interaction with B, A's options for using force to resolve the situation – in a morally justifiable manner – are restricted to either (i) waiting for B's aggression to commence before responding with force, or (ii) attempting to apprehend members of B's forces, and thereby initiating a combat situation themselves. And in either case, it seems probable that *more* preventable killing will eventuate than would have occurred if A simply approached its engagement with B under a combat paradigm, rather than a policing paradigm. Why? Because even if state A's military force enjoys an enormous superiority in combat capability over state B's force, A is very unlikely to be able to effectively police violent wrongdoing by B, if B's members remain ready and willing to 'go to war' with A, before acquiescing to their own arrest. And as the perennial occurrence of asymmetric conflict demonstrates, the Bs of the world often *are* prepared to go to war with the As of the world, even while faced with seemingly insuperable disadvantages.

In short, it seems to me that the idea that large inequalities between armed forces automatically transform combat into policing trades upon a kind of 'rational actor' theory of armed conflict – the key supposition being that actors who can see that they are destined to lose in the event of full-scale combat will allow themselves to be apprehended, before engaging in futile violence for the sake of an unwinnable conflict. But this supposition is unsafe. In an asymmetric conflict where both sides understand their interaction as one of full-scale combat, and act accordingly, it seems more likely that a decisive outcome will be achieved quickly, with most of the costs being incurred by the disadvantaged state. By contrast, where the Superpower in such a conflict sees itself as policing violent wrongdoing, while the Underdog thinks of itself as fighting a war, the conflict is more likely to be drawn out, and significant human costs are more likely to be incurred on both sides. And (particularly in cases where the Underdog is an unjust aggressor) the first scenario clearly seems like the lesser of two evils. If that is right, then in these sorts of scenarios, it seems like a significant mistake for a Superpower to approach its conflict with an Underdog under the auspices of a policing paradigm.

Kahn and others are right to insist on the ethical indefensibility of the Superpower annihilating Underdog forces *en masse*. But where complete disengagement is not a viable option either – where the circumstances of the conflict *necessitate* an active response – the question we have to ask is: how *should* the technologically advantaged state conduct itself? And my point is that we cannot expect the military superpowers of the world to eschew full-scale combat in favour of policing, not if by policing we mean anything like what we normally mean by the term in domestic political contexts. 'If combatants are no longer a threat', Kahn says, then 'they are no more appropriate targets than non-combatants' (2002, p. 5). I am not objecting to *this* claim – rather, I am arguing that it is insufficient to establish Kahn's claims about the Superpower's duty to adopt a policing approach in all contexts. In the kind of asymmetric conflicts we are considering, wherever it *is* the case that the Underdog's personnel pose little or no threat to the Superpower's personnel, this is the case *only while* (and only because) the Superpower is actually exploiting its superior military capabilities in order to dominate its opponent in combat. If the Superpower eschews full-scale combat, in favour of a policing approach, it becomes possible once again for the Underdog's personnel to carry out lethal attacks on the Superpower's personnel. As long as the individual soldiers who are responsible for carrying out the Superpower's on-the-ground policing activity remain vulnerable to such attacks, the demand that they abstain from combat is equivalent to a demand that they relinquish their positional advantage, and the relative degree of safety which is concomitant with that advantage, in order to risk death at the hands of their opponents. And especially if the conflict stems from the Underdog's unjust aggression, this demand seems unreasonable. Whatever follows from Kahn's paradox of riskless warfare, then, it cannot be an across-the-board obligation, on the part of Superpowers, to approach political conflicts with Underdogs under a policing paradigm, *instead* of a warfare paradigm.

The Technologically Enhanced Soldier as (Relatively) Invulnerable Police Officer

That is how things currently stand, at any rate. But soldier enhancement technologies have the potential to significantly alter the structure of conflicts between Superpowers and Underdogs – indeed, to transform the circumstances of threat and risk that obtain in these conflicts, in a way that will make it possible for the Superpower to adopt a policing approach in its conflict with the Underdog, but *without* its personnel relinquishing the relatively unthreatened position they would enjoy in full-scale combat with Underdog forces. Obviously not all types of soldier enhancement are pertinent in this connection. But *one* of the core aims of soldier enhancement – for instance, one of the central research agendas pursued by MIT's Institute for Soldier Nanotechnologies (see http://isnweb.mit.edu) – is to adapt revolutionary materials technologies, in order to equip the Superpower's military personnel with body armour and life-support systems, which will render them highly resistant to a wide spectrum of normally lethal physical threats, including projectile ammunition, shockwaves, incendiary agents, neurotoxic agents and vesicant agents. It is possible, naturally, that the promises made on behalf of this technological research agenda are exaggerated. On the other hand, technological developments sometimes outpace expectations, even the ambitious expectations of those undertaking the research. Suffice it to say, there is at least some non-trivial possibility that, in coming decades, US soldiers who are deployed in hostile territory will be equipped with armour and life-support systems which – from the more modestly equipped Underdog military force's perspective – will make the US soldier *extremely* hard to seriously injure, and even harder to kill.[5] To the extent that this transpires, the situation of the US soldier on-the-ground will become much more like the situation of the present-day soldier employing remote weaponry in a combat situation: it is not completely impossible for him to be injured or killed by enemy combatants, but the threat that he poses to the enemy's life drastically outstrips the threat that the enemy poses to his life. What is significant about this prospect, to put it another way, is that it recreates – at a *micro* level (that is, in the up-close interaction between opposing fighters 'on the ground') – the asymmetric dynamic of risk and threat that obtains between the Superpower and Underdog at the *macro* level. At present, the Underdog forces *as a whole* pose only a negligible threat to the Superpower's forces *as a whole*; and that dynamic is preserved for some members of the Superpower's military force (for example Underdog combatants pose, at most, only a negligible threat to the Superpower's fighter pilots and drone operators). Given the development of effective soldier enhancement technology, the prospect is that even soldiers on the ground, in relatively close proximity to hostile enemy combatants, will be in a similar (relatively) unthreatened position.

Under these conditions, Kahn's controversial claim – that Superpowers like the US must, in conflicts with Underdogs, eschew full-scale combat in favour of policing – becomes more plausible. Under these conditions it becomes possible for the Superpower's personnel to carry out the key tasks of policing – for example apprehending wrongdoers, maintaining law and order – without thereby incurring the kind of vulnerability to lethal attacks by enemy combatants, which *would* come along with a shift from a combat footing to policing operations under current conditions. If effective soldier enhancement technologies like those mooted above are achieved, then the Superpower's technologically enhanced troops will be in a position to police the conduct of the Underdog, in a way that the Superpower's troops are not today, notwithstanding the already-existing (macro-level) disparities between Superpowers' and Underdogs' warfighting capabilities.

Obviously Superpowers *already* use their military personnel to engage in police-type activities in international conflicts, most notably in their deployment of peacekeeping forces charged with upholding law and order, whether in post-war zones, or as interventions in ongoing civil wars. One might allow that soldier enhancement technologies make it easier for these policing activities to be safely undertaken, while denying my contention that these technologies trigger an

---

5   The kind of soldier enhancements I am considering here are ones that pertain to the soldier's equipment and accoutrements, rather than his or her basic biological functions, and thus the question of whether a particular enhanced *individual* may justifiably engage in full-scale combat does not really arise. The question is whether soldiers with access to the relevant equipment and accoutrements may justifiably engage in full-scale combat, in cases where their opponents lack these resources.

obligation for Superpowers to more widely eschew combat in favour of policing. If anything, though, the case I have been making will be reinforced by thinking about the parallels here. At present, Superpowers typically only engage in police-like activities in circumstances where their military advantages are great enough that personnel can engage in those activities without being subject to the kind of imminent, reciprocal lethal threats that characterise normal combat. Soldier enhancement technologies of the kind I am adverting to widen the range of cases that can be characterised in those terms, to the point where all (or nearly all) conflicts between Superpowers and Underdogs will be ones in which Superpower personnel are highly invulnerable to lethal attack from their opponents. If that is right, then policing is always (or nearly always) the paradigm of conflict that ought to structure the character of the interaction between Superpower personnel and Underdog personnel in militarised international conflict.

I will finish by stressing what I briefly noted earlier. Much of this discussion is immaterial in cases where the Superpower's conduct, in its conflict with the Underdog, is unjust *ad bellum*. And though I will not argue as much here, it is doubtful that most (or even *many*) conflicts between Superpowers and Underdogs are ones in which the Superpower can assert the justice of its conduct *ad bellum*. If that is right, then what is the upshot of this discussion? The upshot is that even more is required of the Superpower, in order to acquit itself justly in international conflicts, than we might have supposed. It is not sufficient for the Superpower to have just grounds for entering into an armed conflict with an Underdog *ad bellum*. Nor is it sufficient for the Superpower to adhere to the requirements of *jus in bello* within that conflict. The superpower must, in addition to these demands, take on the significant ethical burdens that come with eschewing full-scale combat and, instead, carrying out the duties of law enforcement and abiding by the responsibilities that those duties imply. Soldier enhancement technologies have an ethical impact, in this arena, primarily because they have the potential to remove a key objection to Kahn's contention that it is morally obligatory for the Superpower to reconfigure its approach in this way.[6]

References

Beauchamp, Z. and Savulescu, J. 2013, 'Robot guardians: Teleoperated combat vehicles in humanitarian military intervention', in B.J. Strawser (ed.), *Killing by Remote Control: The Ethics of Unmanned Military*. Oxford: Oxford University Press.

---

6 This chapter builds upon ideas and arguments that I originally developed in a co-authored paper with Robert Sparrow (see Simpson and Sparrow 2014). I should stress, though, that I am solely responsible for any weaknesses in the argument and analysis here. Thanks to Toby Handfield, Rob Sparrow and Ryan Tonkens, and an anonymous referee, for their comments on earlier versions of this chapter.

Coady, C.A.J. 2008, 'The status of combatants', in D. Rodin and H. Shue (eds), *Just and Unjust Warriors: The Moral and Legal Status of Soldiers*. Oxford: Oxford University Press.

Dunlap Jnr, C.J. 1999, 'Technology: recomplicating moral life for the nation's defenders', *Parameters: US Army War College Quarterly*, Autumn, 24–53.

Fabre, C. 2012, *Cosmopolitan War*. Oxford: Oxford University Press.

Galliott, J.C. 2012a, 'Uninhabited aerial vehicles and the asymmetry objection: a response to Strawser', *Journal of Military Ethics*, 11(1): 58–66.

Galliott, J.C. 2012b, 'Closing with completeness: the asymmetric drone warfare debate', *Journal of Military Ethics*, 11(4): 353–6.

Galliott, J.C. 2015, *Military Robots: Mapping the Moral Landscape*. Farnham: Ashgate.

Kahn, P.W. 2002, 'The paradox of riskless warfare', *Philosophy & Public Policy Quarterly*, 22(3): 2–8.

Killmister, S. 2008, 'Remote weaponry: the ethical implications', *Journal of Applied Philosophy*, 25(2): 121–33.

McMahan, J. 2009, *Killing in War*. Oxford: Clarendon Press.

Simpson, R.M. and Sparrow, R. 2014, 'Nanotechnologically enhanced combat systems: the downside of invulnerability', in B. Gordijn and A.M. Cutter (eds), *In Pursuit of Nanoethics: The International Library of Ethics, Law and Technology*, 10. Dordrecht: Springer.

Steinhoff, U. 2013, 'Killing them safely: Extreme asymmetry and its discontents', in B.J. Strawser (ed.), *Killing by Remote Control: The Ethics of Unmanned Military*. Oxford: Oxford University Press.

Strawser, B.J. 2010, 'Moral predators: The duty to employ uninhabited aerial vehicles', *Journal of Military Ethics*, 9(4): 342–68.

Walzer, M. 2006, *Just and Unjust Wars*, 4th Edn. New York: Basic Books.

PART III
Military Medical Ethics

Chapter 8

Among Super Soldiers, Killing Machines and Addicted Soldiers: The Ambivalent Relationship between the Military and Synthetic Drugs[1]

Anke Snoek

The military has a complicated and multilayered relationship with synthetic drugs, which is critically distinct from but highly influenced by the view on drug use in society as a whole. Synthetic drugs were often developed and introduced by the military as a medicine or an enhancement substance. Similarly, soldiers have been exposed to war situations in which both access to drugs and drug use were treated differently than under normal circumstances. Today, the deliberate introduction or tolerated use of synthetic drugs leads to complicated physical and moral dilemmas, including outbreaks of substance dependency (addiction) inside and outside the military or challenges to the cognitive and moral functioning of enhanced soldiers.

In this chapter we will analyse several cases from the American Civil War, the two World Wars and the American Vietnam War, and contemporary research in enhancement substance, to determine how drug use can be analysed and understood in both physical and moral (ethical) terms. This will require a discussion of drug use at different levels. First, we will address the consequences of drug use for the physical and mental sanity of soldiers, during and after wartime, irrespective of the reason for drug use. Second, we will look into the moral questions related to drug use for the enhancement of soldiers, that is, as a method for modern warfare. The moral dimension has at least two different angles: (i) the moral responsibility of superiors administering drugs to their inferiors who are exposed to the rule of full obedience, and (ii) the ethical consequences of enhancement for moral judgement by soldiers in the grey zone between acts of war and war crimes (the difference between the super soldier and the killing machine).

Three key messages emerge from this analysis. The first message reveals that the potential of addiction to synthetic drugs is not a mere consequence of the substance. Rather, it is a combination of substance, personal characteristics and context in which the substance is used. More often than not, one tends to focus on

1   The author would like to thank Ruud van der Helm and Jai Galliott for their feedback.

the properties of substances, rather than the social embedding of substance use. Normative expectations of society tend to play a large role in how drugs are used in the military. With every new war, or every new era, new substances emerge, while simultaneously our definition of addiction, the value we lay on being free of intoxication and dependency, our expectations of soldiers and the purpose of war are evolving as well.

Second, drug use always needs to be considered from a war ethics perspective in which responsibilities for acts of war are to be distinguished from war crimes, and individual responsibilities (the drug using soldier deciding to pull the trigger or not) from hierarchical responsibilities (the superior taking responsibility for the consequences of drug enhanced or drug using soldiers for war crimes committed while in action).

The final message relates to the power relations within the military and the position of the drug user. Drug use is either promoted by superiors as part of enhancement therapies or suppressed by superiors for disciplinary reasons. Soldiers either need to be put on drugs (enhancement) or treated against drugs (addiction), but they are in both cases considered the passive agents in the equation. However, since drug use in war situations is far from one-dimensional, the role of the drug user should be taken more seriously to address both physical and moral dilemmas that arise from it. We tend to overly focus on the properties of a substance rather than listening to the insights of those people taking them, a position that is seriously impeding a thorough approach to understanding the link between military and substance use.

The Emergence of 'The Army Disease'

In 1971, alarming reports about the use of heroin by service men in Vietnam reached the government of the United States. Almost half of the men (45 per cent) had tried narcotics (heroin and/or opium), and 20 per cent reported to be addicted to narcotics. High use of alcohol (92 per cent), marijuana (69 per cent), amphetamines (25 per cent) and barbiturates (23 per cent) were also reported (Robins 1974), but the US government was mainly concerned about the use of narcotics because they are associated with heavy dependency and chronic relapse. The military felt responsible for exposing the service men to heroin by sending them to Vietnam where heroin and opium were easily available. The military probably was also concerned about being responsible for a heroin epidemic in the States.

Since the Civil War, the American military became more aware of the dangers of using narcotics in wartime. Narcotics like opium and morphine made their entry as a medicine in the military during the Civil War. Opium and morphine were used to treat a whole range of diseases like dysentery, diarrhoea, typhus, tetanus, syphilis and headaches. During the Civil War, it was one of the most important medicines used to treat wounded soldiers. After the war, however, many returned soldiers were dependent on morphine and opium, and this dependence on opioids became retrospectively known as 'the army disease' and was held responsible for mass drug addiction in America and related crimes (Lewy 2014). Opium and morphine turned out not to be harmless medicine, but a dangerous source of addiction.

The US government, concerned about history repeating itself, designed a detoxification and monitoring program for the GIs from Vietnam. After the war, the men were only allowed to go home after providing a clean urine sample (Operation Golden Flow). If they failed to provide a clean sample, they were offered a detoxification program before returning home. Once returned home, the soldiers were followed up after one year, and after three years to monitor if they were re-addicted again. The results were astonishing. At the first year follow-up, only 5 per cent of those who had been addicted in Vietnam experienced a relapse while being in the US. Of the control group, existing of addicted civilians who just finished a treatment program in the USA, two-thirds were already re-addicted after six months. At the three-year follow-up, 12 per cent of the veterans had been re-addicted somewhere in the past three years, but their relapse had usually been very brief (Robins 1993). These findings challenged the previous experiences the military had with the use of narcotics, and refuelled the debate on whether narcotics were an evil source of addiction, or an acceptable substance that could be used as a medicine or for leisure in stressful wartimes.

What is Addiction?

What does the above example of dependency on opioids teach us about the way addictive substances work? Lewy (2014) has pointed out that we only relatively recently considered addiction a disease, yet, this idea is now so engrained 'that we cannot imagine drug use without it' (Levy 2014, p. 108). Does substance use automatically lead to addiction? The example of the returned Vietnam soldiers seems to suggest that not all substance use leads to addiction. Although the common view is that addictive substances are bad for people because of some intrinsic properties of that substance, in reality, what makes a substance addictive are not only its properties. Zinberg (1984) has shown that addictive behaviour is always a combination of substance, set and setting. Or in other words, it is an interaction between the properties of a certain drug, and the influence it has on a specific person in a specific context. Let us look closer at these three determinants.

Different drugs have different properties, and are mostly distinguished as stimulants, depressants and hallucinogens. Different substances have different addiction potentials. Although drugs can be globally categorised based on their properties, their effect is not similar on everyone.

Personal characteristics play a huge role in determining the effect of a substance. For example the physical properties of a person: their race, their weight, their brain chemistry, if they have just eaten or not, if they are good sleepers or not and so on, or the psychological properties of a person: if someone is impulsive or sensation seeking, his/her psychological co-morbidity and the reasons why someone takes a substance, all matter. There are many personal differences in the effect of a substance. While amphetamines make some people more active, people with attention deficit disorder are able to concentrate better when they use Ritalin, an amphetamine-based medicine. While some people become more social when they drink heavily, others become aggressive. If you willingly experiment with LSD, it will have a different effect then if someone spikes your drink with it.

Different settings also account for different effects of the substance. A setting can refer to whether a drug is illegal or not, and how you feel about this illegality, or the setting in which you normally use and the anticipation of your body on that setting. If you expect your dessert to be sweet, and it turns out to be sour, you will not recognise the sour taste for a moment and only experience a disgusting taste, although in general you might like sourness very much. The same is the case with substance use. Substance use is often ritualised by the user, and this ritual plays an important role in the body and mind's anticipation on the effect. When we are stressed or in an environment we feel less confident in, substances will have another effect than when we consume them in the safe environment we always use them in. Even when the same amount is consumed, drinking with friends, drinking at a reception at work or drinking alone can have quite different effects, depending on how we feel about the setting.

The setting in which someone uses becomes highly salient with cues. Substance use is often described in terms of conditioned learning (Robinson and Berridge 1993). The Pavlovian dog already starts to drool when he hears the bell preceding his dinner, and most likely he drools already when he enters the room. In the same way, substance-related cues become very salient to the user and evoke a strong craving. The spectrum of these cues is enormous: when an alcoholic enters the room, an empty glass, lid of a can, a fridge, a coaster, a sports game on television and so on will immediately draw attention and can provoke strong cravings. Many addicted people state that getting clean or sober is not the hardest part. Detoxification is relatively easy. However, staying clean day after day is the real challenge. What makes staying clean especially hard is that people often return to the same social setting in which they used, and the setting is drenched with substance-related cues. On top of that, people often lost their job and the trust of their loved ones and sometimes their own believe in self-efficacy. They feel wounded in their self-identity and limited in their goal setting (Snoek et al. 2012; Kennett et al. 2015).

What Vietnam Teaches Us About Addiction

The three factors described by Zinberg – the properties of the drug, the characteristics of the person and the setting – are closely intertwined in determining the effect of a substance on a specific person in a specific place. Returning to the Vietnam example with this knowledge, Robins (1974) points out that the personal characteristics of the Vietnam soldiers are quite different from the population that is mostly featured in addiction studies. While the Vietnam soldiers represented the general population, most addiction studies represent a clinical population. People from a clinical population are more likely to have social, financial and mental health problems that are still present when the physical dependency is overcome. The general population, however, is marked by an absence of these risk factors, and spontaneous recovery is more likely to appear in that group. Robins emphasises that the veterans with the heaviest dependency probably belonged to this higher risk group, since they reported higher rates of pre-service antisocial behaviours like fighting, truanting, drunkenness, arrest and school expulsion.

The setting in which the soldiers used seems to be highly important as well in determining the difference in relapse. The soldiers used the opiates in a specific setting: during wartime in Vietnam while serving. This setting is relevant in different ways: they were in a different social context and role, there was peer pressure to use, the use was less stigmatised than at home, they were subjected to stress, physical pain, exhaustion and boredom, the opiates were of a high quality, the substance-related cues that evoke cravings were strongly related to the setting of Vietnam, the soldiers had context-specific reasons to use.

By requiring that the soldiers where clean before returning home, and by offering them treatment, an important disruption of the setting-related conditioned learning was achieved. Soldiers were given a clear signal: it is understandable that you used in Vietnam, it is not acceptable that you will use at home. On top of that, soldiers were given a huge incentive to stop: they could only return home when clean. By giving them a chance to get clean on their own, their belief in self-efficacy with regard to their use was enhanced. The identity of the soldiers was not burdened with the addiction stigma. Unlike most other substance-dependent groups, or addicted soldiers in the past,[2] the soldiers were not punished or stigmatised for their use, their dependency was treated as a medical problem and a situational problem (the special circumstances in Vietnam) and not as a moral problem. The follow-ups one and three years later showed people that the military cared about their wellbeing and probably provided people with an extra incentive to stay clean. By making sure the soldiers were not stigmatised and returned

---

2 Lewy (2014) points out that during the Civil War the craving for opioids was not regarded as a medical condition, but rather as a bad habit, like going to brothels and gambling. The returned soldiers who could not control their urge to use opioids were concerned about being sinners or morally weak (Lewy 2014, p. 108). The governmental response to these morally weak persons was consequentially quite different than nowadays. Bergen-Cico (2012) describes how after World War I, there was an 'outbreak of morphinism' among German veterans who received medical treatment during the War. Their morphinism was heavily criminalised, they were reported to a medical board and could be institutionalised indefinitely or even sterilised or euthanised due to the 'Law for the Prevention of Offspring with Hereditary Diseases' (Bergen-Cico 2012, p. 44). Although after the Civil war morphinism was not recognised as a disease, retrospectively, the dependency on morphine and opium was labelled 'the army disease'.

without physical dependency to a new setting probably strongly contributed to the low relapse rate among Vietnam soldiers after they returned home.

From Army Disease to Enhancement Therapies

Robins' (1974) main conclusion is that the addictive properties of heroin are severely overestimated. He states that the 'heroin addict' is a mythological creature. Heroin is not that dangerous and quite safe to use in the right way. Vietnam veterans reported no negative effects of their heroin use both in Vietnam and back home. This is also one of the images the military wanted to convey to the public at large: tolerating heroin use in Vietnam did not mean that superiors exposed their men to an evil drug, but rather to a way for the soldiers to relax in a relatively harmless way, with no lasting negative side effects. In some respect, Robins seemed to work towards a political conclusion: some substances are safe to use in the military, since people can be effectively detoxified, no harm was done and no long-lasting effects will be seen in the community.

Robins was working on his research around the same time that the military began to seriously explore the possibilities of using synthetic drugs in warfare. Although the use of substances in war is probably as old as war itself (Bergen-Cico 2012), the stakes got higher after the Second World War. During World War II, Nazi scientists developed synthetic drugs not only to enhance their own soldiers, but also to incapacitate their enemies. With regard to enhancement, the scientists developed a pill that was supposed to make their soldiers super human, the so-called D-IX pill, containing cocaine, methamphetamine and a morphine-based painkiller (ibid., pp. 40–41). Amphethamines are a stimulant drug that enhance mental focus, feelings of power, strength, energy and confidence, while decreasing the need for food or sleep. With regard to incapacitating their enemies, the Nazi scientists experimented with hallucinogens to see if these could be used for mind control of the enemy. These experiments proved to be less successful: instead of being able to control the minds of prisoners on hallucinogens, the prisoners were no longer able to perform the simplest tasks. Back then, very little was known about the effects of different drugs, or the results of the experiments of the Nazis. But the technology to produce synthetic drugs evolved rapidly, such that the creation and use of synthetic drugs was seen as a promising tool in warfare. This was intensified during the Cold War, when the American military increasingly experimented with the development of synthetic drugs, afraid that the Russians might have gained access to the outcomes of the Nazi experiments.

The Creation of Super Soldiers

The maturing of synthetic drugs provided many possibilities to enhance soldiers, that is, to deliberately administer drugs that allow a soldier to perform better and longer under more extreme circumstances. The enhancement is part and parcel of the increasing role of high technology in warfare. The soldier is no longer a man carrying a gun who is sent into battle. Rather, he has become part of a complicated technological complex in which he has to master advanced technological devices just as much as the devices enhance the soldiers' capabilities. Drug enhancement is simply one aspect of many technological inventions to create super soldiers. However, enhancement is far from problematic in both physical and ethical terms. As we saw with the Vietnam example, many variables determine what effect a substance has on a specific person in a specific setting. These variables constantly change, including, and arguably most importantly, the moral responsibilities of soldiers. Below we will analyse two trends: (i) the increased institutionalisation of the use of enhancement drugs, and (ii) the suppression of mechanisms of self-regulation by the drug users.

Compulsory Enhancement

During World War I and World War II, it was almost considered cruel to deny soldiers alcohol. The use of alcohol was seen as a necessary coping mechanism for soldiers facing the horrors of the battlefield (Bergen-Cico 2012). The public opinion on substance use by soldiers changed radically during the Vietnam War. Substance use was held responsible for the internal collapse of the armed forces due to lack of discipline among the soldiers. The newspapers were dominated by stories of how stoned soldiers fired at their own people, and how the Vietnamese sold opioids to the soldiers to make them less capable of doing their jobs. In short, the 'myth of the addicted army' was born (Kuzmarov 2009). Although Robins (1974) provided evidence that the soldiers used the opioids in a relatively safe way, and that they were enhancing rather than impairing the soldiers' capacities, the liberal view on the advantages of self-medication had irrevocably changed. Where in earlier wars the sheer quantity of the soldiers was most important to outnumber the enemy, in modern warfare soldiers became highly trained professionals who are not supposed to fill the trenches, but to make advanced technical and moral judgements under stressful conditions.

As a consequence, the prescribed use of substances has taken flight. Allenby and Sarewitz (2011) remark that the most enhanced people in our society today are soldiers. Soldiers can simply not afford not to use enhancing substances, given the responsibilities they have for the lives they are protecting, including their own lives. Added to that, they are exposed to high-stress situations, leading to significantly higher incidence of Post-Traumatic Stress Disorder (PTSD) among war veterans (we will return to this latter issue later).

However, drug use for enhancement is still surrounded by ambivalence. This is well reflected in the policies around the use of Dexedrine (dextroamphetamine) among jet pilots to help them stay alert during long missions. Formally, the use of Dexedrine is voluntary, and American pilots are asked to sign a consent form before they get access to the enhancement drug. However, the form also notes that pilots can be grounded if they decline, which could have serious implications for their careers. As a consequence, the consumption of Dexedrine by military pilots is by any standard compulsory rather than voluntary. Whereas the military is rightly expected to regulate experimental substance use due to the health risks associated with it, following Zinberg's model that strongly emphasises the role of the setting, it is also taking serious risks making the use of the enhancement substance compulsory.

The Suppression of Self-regulation

The institutionalisation of drug use goes hand in hand with another trend: the suppression of mechanisms of self-regulation by drug users. Whereas the administering of drugs by the military is associated with the safe enhancement of soldiers to make them physically and cognitively stronger, when it is the individual soldier who takes the initiative to use a substance, it is increasingly associated with addictive, drug-seeking behaviour and intoxification, which maps into the myth of the addictive army. An interesting case that reached the newspapers[3] took place in September 2013, when 17 British soldiers, including two sergeant majors, were charged with the use of the performance-enhancing substance Ephedrine. Paradoxically, Ephedrine is closely related to the amphetamine the military uses to enhance its soldiers. Although the culprits were described as 'gym rats', rather than drug addicts, they were all discharged from the military.

This suppression of self-regulation in favour of institutionalised administering is a crucial issue in view of the safe enhancement of individual soldiers. Most discussions so far have focused on the question whether medical ethics in time of war are essentially different from general medical ethics or not,[4] and if soldiers should be asked for their consent or that the decision is part of a higher goal rather than the rights of the individual (Gross 2004) (See also Chapter 4 in this volume on informed consent). But with regard to the use of substances as enhancement, the discussion should rather focus on what is more effective and safe. For that, prejudices and fears regarding self-regulation need to be replaced by sound analytical work on the intricacies of institutionalised drug administration. A first distinction to let go off is the one between drug use as related to moral weakness (addiction) and drug use for physical enhancement. To understand this critical point, enhancement needs to be repositioned in its ethical context.

---

3 http://www.dailymail.co.uk/news/article-2420904/British-Army-drug-scandal-17-soldiers-caught-using-performance-enhancing-chemicals.html

4 The World Medical Association states that they should be the same, while Gross (2004) argues that they are essentially different.

Super Soldiers, Ethics and Responsibility

On 16 April 2002, two American Air National Guard pilots flying over Tarnak Farms in Afghanistan believed they saw surface-to-air fire. They contacted the flight control and asked permission to drop a bomb. However, feeling threatened, 35 seconds later the pilots did effectively drop the bomb, stating that they acted out of self-defence. Soon after that, the answer of the flight control came, stating they were 'friendlies', Canadian soldiers conducting a firing exercise. This warning came too late, four Canadian soldiers were killed on the spot and eight wounded. As Annas and Annas (2009) analysed, much went wrong in this situation. The pilots were not briefed in advance about the presence of these 'friendlies'. Instead, their briefing focused on warnings about Taliban ambushes and surface-to-air fighting in that area, and on cruelties committed by the Taliban. One of the salient points the trial focused on was the fact that the pilots had been administered Dexedrine, which may have impaired their judgement. Ironically the prescribing information for Dexedrine reads: 'Amphetamines may impair the ability of the patient to engage in potentially hazardous activities such as operating machinery or vehicles' (Annas and Annas 2009, p. 293). That notwithstanding, the final court ruling read that the pilots were considered responsible and they were reprimanded. However, Annas and Annas (2009) suggest that they had been scapegoated in order to defend the military's enhancement policy. As such, they raise a fundamental question regarding ultimate responsibility for behaviour conducted under influence of an enhancement substance.

The obvious ambivalence resides in the level to which the absence of a free choice in taking the enhancement substance can be used as a justification for war acts 'under influence' with unintended consequences, whether these are friendlies or outright war crimes. As we discussed earlier, the effect a substance provokes hugely varies from person to person, and from setting to setting. In that sense, compulsory enhancement can work counterproductive when compared to self-regulation under medical supervision. The pilots argued that their judgement was impaired due to the compulsory use of Dexedrine, and if they would have been able to regulate their Dexedrine use themselves, that they would have judged differently when not under the influence of this so called 'go pill'. Substance use literature strongly advises to take such an experience very seriously, irrespective of the claim on whether this is a real effect of the drug, or an imagined one. Had the pilots have been given a choice in the use of Dexedrine, they would probably have taken the drug anyway; they would arguably have come to the same decision, but they would feel less victimised by the outcome of the whole tragedy, and it would instate an ethical foundation based on which acts of war can be judged and tried.

The most pressing ethical question therefore is how we define 'super soldiers', and which properties of soldiers do we want to enhance (see also Chapter 4, Bio-Technological Challenges to Autonomy and Chapter 6, enhancing responsibility). The nature of modern warfare has changed and combines different objectives or rationales beyond simply defending a territory. Regime change and humanitarian imperatives are intricately linked in with defence of strategic interests. In particular the humanitarian imperative of warfare has also changed our perception of soldiers, who become its moral agents rather than killing machines. In that sense a drug that blunts moral responses to horrific events seems to surpass the military's goals, or, in the words of The President's Council on Bioethics, we should do everything to ensure that 'men remain human even in moments of great crisis' (2003, p. 152).

Super Soldiers versus Killing Machines

Besides judgement errors in the act of war, a fundamental ethical risk of drugs is the one related to blunting memory. Having truthful memories in a combat situation is not only a personal matter, but it entails a responsibility towards the world to bear witness and testify. Bergen-Cico (2012) very clearly states that 'the use of a drug to block soldiers' emotional response to killing threatens the ethical basis that preserves humanity' (p. 13). In particular the fact that Post-Traumatic Stress Disorder is a significant cause of morbidity in military personnel, and hugely influences the quality of life of soldiers and their families (Searcy et al. 2012), warrants for caution in the use of memory blunting drugs.

Unsurprisingly, the prevalence of Post-Traumatic Stress Disorder among soldiers is higher than among the general population, with a life-time prevalence of 30 per cent compared to 8 per cent in the general population (Johnson et al. 2009). PTSD is associated with self-medication with alcohol and other substances, often leading to addiction. Approximately 75 per cent of military personnel diagnosed with PTSD have a co-occurring substance abuse disorder. PTSD is also associated with a higher suicide risk and violent outbursts (Searcy et al. 2012). Although there are psychological interventions available, in some cases PTSD is very hard to treat. Recently there are some promising results in the development of pharmacological treatments that could prevent the occurrence of PTSD. Let us first look at how PTSD emerges from a neuro-scientific point of view.

We do not remember all the events that happen to us equally well. How strong and durable an event will be engraved in our memory depends on the amount of adrenaline that is released during the events. Feeling of euphoria, pleasure, fear, horror and stress all release adrenaline into our system. In that way we remember the things that are important to us for our happiness or avoiding dangerous situations. When we remember an event, we also remember the emotion we felt at the time of the experience. In PTSD a traumatic event causes a release of such a high level of adrenaline that the memory encoding system becomes overactive. The memory becomes strong and persistent. The memory is easily relived, triggered by seamlessly harmless cues from the environment, like unsuspected noises or fireworks. Every time the memory is relived, this is accompanied by the original feelings of stress, which again release a high amount of adrenaline, which results in the new event also being engraved in the memory. Having PTSD can make daily life extremely hard or even impossible (The President's Council on Bioethics 2003).

Medical drugs like corticosteroids and propranolol that block the release of adrenaline can influence the memory encoding systems. When taken shortly after the traumatic event occurred, these medicines can blunt the memory and detach the memory from the original emotion. These medicines however do not influence our long-term memory, and are only effective when taken soon after the traumatic event. The fundamental difficulty is that not every traumatic event leads to PTSD, and that the medical drug should be taken before PTSD can even be diagnosed (The President's Council on Bioethics 2003). It is very hard to predict beforehand who will develop PTSD after experiencing violent combat. This leaves us with the dilemma who should be administered the medicine, and if the burden PTSD puts on some of the soldiers outweighs the risks of administrating the medicine to a large group of people. The President's Council on Bioethics (2003) emphasises several risks of preventatively administering this kind of medicine to a large group of people. For some people this might not be a medicine, but rather a poison that will make them morally indifferent. They will remember the horrors of the battlefield without any emotion. 'By "rewriting" memories pharmacologically we might succeed in easing real suffering at the risk of falsifying our perception of the world and undermining our true identity"' (ibid., p. 225). Right here, super soldiers are turning into killing machines.

Conclusion

Since the relation between drug use and the military is likely to remain ambivalent, it is essential to permanently and critically analyse and debate the modalities and the consequences of substance use, both on physical and on ethical grounds. As we have discussed, there is no clear-cut answer to whether a drug is effective or desirable, since any substance will need to be assessed against set and setting. How we view a certain substance is highly influenced by our normative evaluations and the standards in society. As we have seen in the above analysis, these standards and normative evaluations constantly change. Today, the trends are towards increased institutional experimentation and against self-regulated use, which reflects today's public opinion which is less tolerant towards intoxication and doping, while there is increasing pressure to perform at levels that cannot be sustained without enhancing substances (a trend visible throughout society and not confined to the military alone). We tend to have negative associations with self-regulated drug use but embrace medicines and enhancement.

However, our negative association with self-regulated substance use makes us ignorant of an important aspect that determines the successfulness of a medicine or an enhancement substance: the personal experience of the user and his/her involvement in determining the objective and context of substance use. Zinberg's work shows that there is good evidence to suggest that enhancement will be more effective when not compulsory. This would imply that the military should take an interest in being more tolerant regarding self-experimenting with substances, provided that it can take place under medical supervision and the results can be used to monitor the effects of substances.

In order for this change to happen, we should let go of the moral condemnation of experimenting with drugs, and our blind trust in prescribed medicine or enhancement substances. We should acknowledge that there is only a thin line between a beneficial medicine, enhancement substance, drug and the poisoning effects of medicine, the impairing effects of enhancements and the addictive effects of drugs. Simultaneously, we should acknowledge that there is another thin line between super soldiers and killing machines, between enhanced performance and blunted emotions. For the old Greeks, the word medicine had a double meaning: healing and poisoning (Lewy 2014). Whether a substance is poisoning or healing does not depend on the substance itself, but to an important extent also on the person who takes it and the setting.

Therefore, it is surprising that what is missing in this debate are the voices of the soldiers themselves in a genuine discussion on how much individual free choice should be accommodated, against broader physical and ethical standards. Although issues of informed consent may be different in this situation than in normal life, this does not mean that soldiers as primary users should remain passive agents. It would put to the fore questions related to which substances do provide what benefits under what circumstances. Or how do they see the relationship between being protected against the horrors of PTSD and the responsibility to have truthful memories and appropriate moral emotions to horrific events; or on the difference between memory blunting with alcohol or marijuana and with propranolol; or regarding which characteristics of themselves would they like to enhance to be able to perform better at their jobs, and in which properties do they want to remain 'human'? Which drugs do they see as an enhancement, a medicine or a scourge? These insights are largely missing in the debate so far.

It would go too far to state that the best judge about the effect of a substance is the person itself, but controlled self-medicating or self-enhancement is likely to be the most promising route to dealing with the multilayered consequences of its use. That is what we seem to expect from our super soldiers nowadays, that they are mentally and physically in control of themselves. This could be reached paradoxically by their self-experimenting with substances under medical supervision, despite the fact that this is commonly associated with morally condemnable addictive behaviour.

References

Allenby, B.R. and Sarewitz, D., 2011. *The Techno-Human Condition*. Cambridge, MA: MIT Press.

Annas, C.L. and Annas, G., 2009. Enhancing the fighting force. *J Contemp Health Law Policy*, 25(2): 283–308.

Bergen-Cico, D.K., 2012. *War and Drugs: The Role of Military Conflict in the Development of Substance Abuse*. Boulder, CO: Paradigm Publishers.

Gross, M.L., 2004. Bioethics and armed conflict: Mapping the moral dimensions of medicine and war. *Hastings Centre Report*, 34(6): 22–30.

Johnson, J., Maxwell, A. and Galea, S., 2009. The epidemiology of posttraumatic stress disorder. *Ann Psychiatry*, 39(6): 326–34.

Kennett J, Vincent N.A. and Snoek A. 2015. Drug addiction and ciminal responsibility. In N Levy and J Clausen (eds) *Handbook on Neuroethics*. The Netherlands: Springer, pp 1065–83.

Kuzmarov, J., 2009. *The Myth of the Addicted Army: Vietnam and the Modern War on Drugs*. Massachusetts: University of Massachusetts Press.

Lewy, J., 2014. The Army Disease: Drug Addiction and the Civil War. *War in History*, 21(1): 102–19.

Robins, L., 1974. *The Vietnam Drug User Returns*. Washington DC US Govt Print. Off.

Robins, L., 1993. Vietnam veterans' rapid recovery from heroin addiction: A fluke or normal expectation? *Addiction*, 88(8), 1041–54.

Robinson, T.E. and Berridge, K.C., 1993. The neural basis of drug craving: An incentive-sensitization theory of addiction. *Brain Research Reviews*, 18(3): 247–91.

Searcy, C.P. et al., 2012. Pharmacological prevention of combat-related PTSD: A literature review. *Military Medicine*, 177(6): 649–54.

Snoek, A., Kennett, J. and Fry, C., 2012. Beyond Dualism: A Plea for an Extended Taxonomy of Agency Impairment in Addiction. *AJOB Neuroscience*, 3(2): 56–7.

The President's Council on Bioethics, 2003. *Beyond Therapy: Biotechnology and the Pursuit of Happiness*. Washington DC.

Zinberg, N.E., 1984. *Drug, Set, and Setting: The Basis for Controlled Intoxicant Use*. New Haven, CT: Yale University Press.

Chapter 9
Bio-Technical Challenges
to Moral Autonomy

Steve Matthews

When are soldier enhancements permissible in so far as they affect moral autonomy?[1] I answer that question by setting out an important condition for moral autonomy: the capacity agents have for psychologically appropriating actions and experiences into a unified morally coherent self-conception.[2] For example, a soldier who, in the course of a mission, recklessly kills an unarmed enemy civilian must take responsibility for his actions. What if drugs were available which erased his memory, and removed all guilt for such an act? If this were to take place, the loss of memory would bring about a loss in the ability to identify with it as his own, and so he would be prevented from even the possibility of attempting to justify to himself or others what he has done. In a sense, such an enhancement is the equivalent of hiding evidence, and in that way it prevents the possibility of any restorative moral process. Such an enhancement has also rendered this individual unable to discharge the obligations that relate to the principles which give him his moral identity as a professional soldier.

I assume there is agreement that memory deletion in such cases corrodes moral autonomy, but rather than painstakingly go through many different examples of enhancement, the strategy here is to put forward a *general* test of what counts. A key question is this: does the enhancement promote in the agent a capacity for responding to reasons that enable morally unified agency, or does it disrupt this capacity? My approach is to begin with moral autonomy as a non-negotiable value, and so if a certain kind of enhancement is a threat to that value, as we just saw above, then it must be rejected on moral grounds. But there may be examples

---

1   I will sometimes use the word 'autonomy' unqualified. It will refer to *moral* autonomy, unless the context suggests otherwise.

2   Those familiar with Christine Korsgaard's work (particularly (2009) will recognise that the theoretical apparatus of this chapter is hers; I make no claims to it. I use it here because (of course) I think it is largely correct, and second because its elegance is well suited to application to the empirical case at hand. I will not talk about authenticity, because I do not think that concept quite hits the mark for moral autonomy. For an excellent discussion of authenticity and autonomy in legal cases see Bublitz and Merkel (2009). They too eschew an authenticity condition in favour of the idea of an agent's capacity for identification and control.

in which, on the contrary, moral autonomy benefits from an enhancement. It would benefit if it helped the agent to avoid losses to moral autonomy the agent *would have* sustained without the enhancement. (I am not claiming the benefits of enhancement stem from causing the agent to have a happier life, even though that too is of course desirable.) Thus, it is an open empirical question – a negotiable value let us say – concerning which enhancements are permissible, just because of the heterogeneous nature of the effects. So, the moral position is not monolithically opposed to enhancing, nor is it too liberal.

Two Assumptions Noted

First, Lin et al. (2013: 21) provide a comprehensive survey of the technologies for warfighter enhancement.[3] They divide enhancements into physical capabilities (such as strength or mobility), cognitive capabilities (such as awareness, attention, memory), the senses (such as sight or smell), metabolism (for example, to improve endurance or absorption) and a miscellany of dual-use research applications (for example, in stress management, bio-resistance to toxins and many others). An interesting question thus pertains to which of these bio-enhancement types is relevant to questions of autonomy. To avoid an endless trawl through such a long list, I will instead focus on a type of enhancement in which there are *salient* effects on moral self-identity. For example, as signposted already, I will be particularly concerned with the use of drug enhancements that affect memory, given the close connection between memory and identity. Indeed I think there is good reason quite generally to focus on the enhancement of cognitive capabilities because of the *direct* connection borne to personhood. That is not to say that improving a person's speed, sight or endurance is not relevant to the question of autonomy, just that those effects do not have a logical link to our central question, whereas changes to such traits as memory do.

Second, this chapter will not address questions arising from the study of war and its conduct.[4] Yet, on my account, the reasons that underlie a threat to moral autonomy do connect (at least in a contingent way) to the conduct of war. For, to put it in Kantian terms, in the formulation of a maxim relating to an act performed in wartime, considerations concerning the justice of one's cause, or just conduct in war, ought to inform this maxim. Assuming just cause, a fully morally autonomous soldier would be cognisant of what is forbidden and what is permitted in the conduct of war. (Obviously going through this intellectual process in the

---

3   Their survey is US-focused and based on 'open-source or unclassified information' (p. 21).

4   Just war theory has roots in work done going back to Augustine, and then more systematically by Aquinas in the thirteenth century. See Augustine (1958), and Aquinas (1948), in the Second Part, Article 1, of the *Summa*. For a modern treatment see Walzer (2004).

heat of battle is not what is envisaged, but rather, these moral rules are instilled during training.[5]) Thus an enhancement that prevents an agent from discharging moral obligations falling out of the considerations of just conduct in war would, *ipso facto*, be *prima facie* impermissible.

The Argument Illustrated

Christine Korsgaard (most recently (2009)) has used the language of self-constitution to characterise the link between moral action and identity.[6] A practical identity is forged through *choosing* and following principles that unify one's agency. The morally autonomous being is one who gives herself laws that are rationally informed by the categorical imperative, and who behaves accordingly. Thus, such actions – that is, acts done for the sake of certain principled ends – performed in an ongoing way, sustain one's personhood; agency itself is the mechanism by which one's personal identity is constructed.[7] On this Kantian conception, agential unity is necessary for moral personhood; let us say that this represents one's core moral identity. Now, human beings have, besides this core identity, contingent professional identities related to their roles (for example, doctor, lawyer or in our case, soldier).[8] So, it is one thing for a human being to globally lose the ability to act on the moral law (occasioning loss of moral personhood itself), but it is another to fail to act on one's professional principles, or not in accordance with an adopted

---

5   Kant addressed this specific question directly. The public official, the officer, the citizen, even the Clergyman, he states, must adhere with obedience to the requirements of their related institutions. They are permitted *publicly* to question the justice of a practice, but privately while on duty, Kant is explicit: despite recognising an order as being inappropriate or useless, the officer 'must obey' (in 'An Answer to the Question: What is Enlightenment? (1784)').

6   In previous work Jeanette Kennett and I used the unity of agency thesis in providing analysis of the losses that attend certain types of psychopathology, for example, in dissociative conditions. See, for example, our 2003.

7   Korsgaard (2009: 12) defines actions as acts-for-the-sake-of-ends, so an action is temporally extended to build in the reason that led to the behaviour.

8   Another contrast is between contingent practical identities *generally* and a core moral identity. The category of contingent practical identities subsumes professional roles in addition to other roles, such as husband, wife, child, parent or friend. My claims here about professional roles should not be taken to extend to these other roles. In general a professional role, more than, say, a close friendship, should dissolve more quickly upon pressure from morality. I do not claim that being a friend puts you beyond the reach of morality, only that the value of close friendship (or parenthood or …) differs in its particulars to certain professions, and so the range of life's activities where application of the categorical imperative should remain dormant is *sui generis*, and typically more extensive. (This has an implication: professional relationships that double as friendships are potentially morally perilous.)

role. This is because you *take on* the role of lawyer, doctor or soldier. And so, as deeply ingrained as the soldier identity may become to you, it is still something you may repudiate.[9]

In professional practice one constitutes an identity through adopting, and acting on, the principles of one's chosen occupation. The normative sources for soldier identity – including especially the principles informing practice – represent the internal standards required just to *be* a soldier. What this demonstrates is that there a logical link between principled action and a practical identity. Moreover, unless one can discharge the principled obligations belonging to his/her identity – as might occur in certain kinds of enhancement rendering a person into a mere tool of armed conflict – s/he cannot govern her/himself under that description. In more familiar sounding language, we might say of such a soldier, 'well, he didn't sign up for *that*'.

Given the sense of autonomy just outlined – involving moral unification based on self-understanding – we should consider the points of tension generated when the agent is attempting to reconcile the demands of morality with those of being a soldier. For example, how might enhancement affect one's ability to forge an autonomous moral self when there is a professional requirement to obey a morally dubious order? On the face of it, drug enhancements that blunted the soldier's capacity to question the order based on a morally knowledgeable response would constitute an attack on moral autonomy. Tensions can also arise in contexts in which a person joins the armed services to become a soldier, then returns to civilian life, and who then suffers an identity crisis generated by the different moral demands of each form of life. This person might ask, 'who am I really?' and in asking this question, the person expresses doubt in relation to his/her capacity to unify him/herself. However, we should *not* state the problem in terms of how different identities are grafted together. Rather, *at any time*, a person must choose to act on principles – professional or private – measured against the universal demands of morality. The tension is always between heteronomously imposed demands (from say my commanding officer) and autonomously chosen courses of behaviour of moral worth.

Let me pause to provide a further argument for rejecting the position I just described, viz., the position that posits an individual as a series of identities grafted together one after the other, testing the question of enhancement and autonomy against that moral metaphysic. What this view does is to artificially dissociate soldiering principles and civilian principles; so it abandons the moral demand for unity. Jonathan Glover (1988: 23) discusses an extreme form of this, referring to the work of Robert Jay Lifton who did a study of Auschwitz doctors. How would a person who conceives of himself as a healer make room for the atrocities he commits as a killer? Lifton answers: by *doubling*. This involves the ' ... creation

---

9   Korsgaard (2009: 23) puts the point even more strongly, saying, ' ... you can walk out even on a factually grounded identity like being a certain person's child ... '.

of two relatively autonomous selves … '.[10] The 'Auschwitz self' operated as an agent for the monstrous deeds of the regime, but partitioned off from a prior normal self. When the doctor (a relatively normal German person) entered Auschwitz life, a metamorphosis occurred to ensure action as Nazi operative, functionally and emotionally independent, of the norms governing ordinary civil life. In this extreme case, the doubling manoeuvre can be pulled off only if the overall individual person containing these morally opposed personalities compartmentalises the narratives belonging to each, and ensures no mixing of them. For to do so – to gain overall moral autonomy – this individual would, quite simply, morally self-destruct. The autonomy I advocate here – if properly activated – would make impossible such extremes of moral doubling.

Cases of enhancements that blunt accountability and increase obedience are obviously a threat to moral autonomy. Let us now consider a case where things are less clear. Army Maj. John Prior was in charge of an infantry company during the Iraq invasion. Several years after returning home he told philosopher Nancy Sherman that even after those years away from the battle zone he endured a struggle with his conscience, as he told her, a struggle involving his 'own personal guilt' (Sherman, 2011: 90). The occasion for this guilt was not an act he regarded intellectually as blameworthy, as attracting moral responsibility. Rather, the guilt stemmed from an accident in which a misfiring gun killed one of his privates. Now let us suppose that following this otherwise severely traumatising accident, John Prior, along with the other witnesses to the tragedy, were administered a newly formulated memory-deleting drug. The effect of the drug was essentially to erase the experiential memory – along with its trauma-inducing features – of the accident. The drug's purpose was to mitigate or eliminate PTSD. In this altered version of the story, let us assume John Prior is later debriefed and told the facts of the accident. His natural human response is to express sadness and deep regret in relation to what has happened, but let us suppose that under these conditions he is largely negatively unaffected by it. In particular, he feels no personal guilt at all.[11]

In this altered version of the story we can ask: what is the effect of the drug on the soldier's capacity for moral self-conception, and so, his moral autonomy? One answer might be that the effect is to hobble it, and that is because the person-over-time who is the soldier-civilian has lost the ability to appropriate his wartime experience, and perhaps, in a Stoic frame of mind, to then work through it, and finally to overcome the emotional dysregulation causing him so much self-accusation.

---

10   Re-quoted in Glover (1988: 23). See Lifton (1986).

11   This is hypothetical speculation, of course. It is an open question what might happen in such a case. Perhaps the real John Prior would be devastated by such news revealed through debriefing. However, I take it as highly plausible that in many cases at least, erasing a potentially trauma-inducing memory would eliminate or reduce personal guilt.

An opposing opinion might be that the drug enhances moral autonomy. It would do so by preventing a range of post-traumatic effects. As someone free of those effects, the soldier-civilian would be less damaged, and so would retain the psychological 'equipment' needed for moral autonomy into the future.

Which answer is right? Neither seems clearly correct, for let us not forget that John Prior bears no moral responsibility for what happened. Autonomy is a matter of governing oneself as a moral agent by drawing on a unified moral self-conception. In this instance the individual is attempting to take his wartime soldiering experiences and unify them with his civilian self-conception, and to do so under *one* set of moral principles. One might argue that Prior's moral regrets for what happened deserve incorporation in his unified moral self-conception, if he is really to sort out who he is morally. But, it is eminently debatable, a genuinely difficult case. It seems unlikely that the nuanced particulars of any situation like this would not be strong enough to shift our views on the matter either way.

In any case, the point of drawing attention to this situation was to show that enhancements can indeed go either way, sometimes improving autonomy, sometimes undermining it. Moreover, the theoretical picture can now be seen in operation. For if we do regard respect for the moral autonomy of the soldier-civilian as indispensable in decisions concerning the effects of enhancement on wartime fighters, we have a moral test for their permissibility.

Autonomy, Agency, Integrity

In this section some expansion of the notion of autonomy is given, especially as it relates to questions about the choices and actions of agents, and what the person of integrity should decide, when it comes to decisions for enhancement. I begin with three background considerations.

First, we can ask whether moral autonomy is of value as a question internal to a community of moral actors – with no concern for those they engage with outside the community – or as a question in which everyone (in related space and time) has the potential to be autonomous. Call the first *context-bound* autonomy, and the second *unrestricted* autonomy. Political and (national) legal systems operate with a notion of context-bound autonomy; morality with the second, as we are.

Second, autonomy (personal or moral) is enabled because it is both a *capacity* and an *achievement*. On my account both are needed. A person may be in possession of an autonomous capacity, but never act on it, whereas the person who acts on a similar capacity may have a range of achievements. Without the capacity, there *cannot* be the achievement (of course!); unless the capacity is put into service, there *will not* be an achievement. Suppose student Alice has a great gift for mathematics. She vaguely realises she has this ability but does not pursue it; her teachers encourage her into food technology instead. Her capacity for mathematics is present but unrealised and the importance of this to her intellectual autonomy is missed.

Third, autonomy (personal or moral) is a matter of degree, not something a person either possesses or not. Think again of Alice. Her unrealised mathematical accomplishment diminishes her autonomy in one aspect of her life. In other aspects, let us suppose, she excels. Perhaps what she learns in food technology leads to a career as a highly respected chef, and so as a career person she governs her life with significant independence. But other areas may be compromised. So she may be able to control her workspace, but let us suppose she is powerless in the domestic environment; her waning health might compromise her capacity for sports, but her intellect remains intact and she competes in a local bridge club; local laws impinge her control over where she may park her car, but her choice of vehicle remains open, and so on. In innumerable ways, the capacity to govern her life the way she sees fit may be enhanced or diminished across different dimensions. If this is right, then we should recognise that enhancements in professional life also impinge moral autonomy in different ways and so in different degrees.

Providing a simple and complete analysis of autonomy, as any honest philosopher will grant, is highly vexed, to say the least. Notwithstanding this, we do have two clear lexical definitions that we may start with: self-determination and self-government. In the first I am the cause of my own behaviour (independent of external causes); in the second, I have 'sovereign' status, and my behaviour tracks laws I make for myself. One might think that self-determination, though true of us, is not true of us in any particularly unique way. There is a perfectly good sense in which some of the cognitively sophisticated animals determine their own behaviour. But, it is highly plausible that, to the best of our knowledge, human beings are the only species whose members (or some of them) give reverence to, and act upon, the moral law they find within. Since this is so, an account of *moral* autonomy would appear better suited to the self-government approach: I constitute myself over time by following a kind of internal constitution, a tenet of which situates me as one among many other rational beings testing their motivations against the universal moral law. As the preceding paragraphs suggest, I am more inclined to support this deeper account of autonomy in which human agents are not only the self-conscious causes of their behaviour, but come to be so because they reflect on, and then choose (again, perhaps only implicitly) to perform act A over B, because the principle behind A has universalisable content. As such, A can be chosen for a reason. In adopting reasons and principles, and then acting on them, persons construct their own moral personhood. This is what Korsgaard means when she says (2009: xii) that the function of action is self-constitution.

According to Korsgaard, the picture we have (ideally) is of a rational agent who in the course of life is deciding which of the practical identities – of say civilian, patriot, soldier and so on – is to be his role. Of course, many of these are held simultaneously. In taking on these identities, he thereby takes on the conventions, rules or professional norms associated with them. But, as Korsgaard points out (p. 24), these identities are *contingently* associated with this rational agent's pre-eminent moral identity. By acting on an order to shoot, the Private responds to the imperative 'obey orders from my Sergeant', and endorses this role.

As a *moral* agent, there is nothing necessary about the role, and, *if* s/he is indeed a morally autonomous being, s/he may withdraw his assent to it. It follows that any condition that would compromise the agent's capacity to withdraw in this way also compromises moral autonomy.

Put it another way: assume a human being has grown up and reached a threshold for moral decision-making. Then there are three elements to moral autonomy, that is, to the way this human being makes these decisions by choosing the principles that constitute him/her: (1) the agent her/himself (2) the moral law (3) a set of practical identities and their associated norms. For the *moral* agent, acting under (2) is necessary, but acting according to some chosen role under (3) is not. For the choice of role – and what it entails while occupying it – has to be made subject to (2). The norms of professional soldiering – like any profession – are not the final filter before the agent acts.[12] The final filter and morally decisive filter is the moral law, or categorical imperative. If the agent were to dissociate into morally separable identities, then as soldier s/he does not (necessarily) subject her/himself to the moral law.[13] As such s/he can be autonomous within the limits of her/his profession, but her/his autonomy is now context-bound, and that disqualifies him/him from a claim that her/his autonomy is properly orientated in moral space. As noted, this kind of picture is highly problematic when it leads to the moral dissociation – or doubling – described by Lifton.

With this picture in mind, it is easy enough to see what counts as moral integrity: an agent exhibits moral integrity if he or she unifies his/her agency by adopting those practical identities that respect the moral law overall. In particular, s/he should not take on a practical identity that does not respect the moral law at all, and s/he should take on multiple practical identities that best suit his/her ability over time to make them cohere with the moral law. Integrity, in this context, is a quasi-technical notion: you exhibit integrity on those occasions when critically important decisions determine who you are, morally speaking: someone who responds to moral reasons, or someone who turns their back on them. The soldier facing a choice to enhance or not, where enhancement threatens their future ability to make morally informed choices, faces a test of their integrity in this sense. The person of integrity will reject such an enhancement because it prevents him/her from remaining a unified – let us say integrated – moral being. As Korsgaard puts it: 'We owe it to ourselves, to our own humanity, to find some roles that we can fill with integrity and dedication … in acknowledging that, we commit ourselves to the value of our humanity just as such' (pp. 24–5).

There is a corollary of this picture of moral autonomy, and again it is quite straightforward: any institution that does not recognise this understanding of agency and integrity has failed to respect its members as morally autonomous

---

12   As discussed, I limit the claim to professions. In friendship, for example, applying the filter (in *every* case of an act involving a moral question) and siding with morality and not the friend would seem to require one thought too many (in Bernard Williams's (1981) phrase).

13   See Korsgaard, p. 51.

beings. Thus, in the present context, an armed service that imposed a regime of enhancements on its members, the effect of which prevented them from acting with integrity, has failed to operationalise a version of the test set out above.

Finding the Sweet Spot: The Case of Virtual Capabilities

In this final section I discuss the case where soldiers are given virtual capabilities. The motivation for doing so is to illustrate a feature of the theoretical account, that enhancements typically do not render their subjects wholly autonomous or not. Rather, moral autonomy is a matter of degree.

Lin et al. (2013: 23) describe a range of hardware items that act as proxies for warfighters, enabling virtual engagement with the enemy. The obvious benefit is to remove the human soldier from harm's way. As an example, they mention the Avatar program developed by the Defense Advanced Research Projects Agency (DARPA). This program, they say (quoting from DARPA 2012: 123)), seeks to 'develop interfaces and algorithms to enable a soldier to effectively partner with a semi-autonomous bi-pedal machine and allow it to act as the soldier's surrogate'.

Let us assume that the bi-pedal machine is acting just as the flesh and blood soldier would – carrying out missions, engaging the enemy, taking risks, responding to orders and so on. In this situation is there a challenge to moral autonomy? Is there a compromise to an individual's capacity to take such virtual experiences, to view them as his own and then to reconcile them with a proper understanding of the moral law? For example, when a bullet is fired from the machine which is in the field because the soldier sitting comfortably behind the lines pressed a button, what *moral* impact does this have on the soldier? How does this moral impact vary from the counterfactual situation in which that very soldier was on the battlefield exactly where the machine was and fires the same bullet? We might hope that in either case the soldier – the person who caused the death – is able to justify this to himself, and to take responsibility for what has been done. An important part of being able to do that is to experience it in the right way. So, our question boils down to this: do enhancements, such as virtual capabilities, furnish soldiers with the right kinds of experience enabling them to take responsibility for their behaviour? (Do not forget that in my definition of moral autonomy, a loss in the ability to take responsibility is therein a loss of moral autonomy.)

My answer is that the moral impact of experience strongly depends on the quality of the technology to imitate real-world environments in relation to the psychological effects of *remote* agency. We thus have a continuum: a technology that simulated an external environment perfectly would theoretically have an invariant moral impact on the soldier (not varying between reality and simulation); as this simulation degraded, so would the moral impact. In the case where the experience is utterly unlike the real experience – the so-called computer game scenario of warfighting – the moral impact would dissociate from it. Thus, virtual capability enhancements of this nature challenge autonomy to an extent commensurate with their authenticity. In degrading the quality of the experience, they diminish its moral impact, and render the soldier less morally autonomous. I think in general this is correct. But let me finish with a qualification.

My claim cannot really be that virtual simulation must be so perfect that the soldier cannot even tell reality from non-reality. In other words it has to build in an *insight* condition. Compare this to aviation simulators. The ideal simulator in that context would be one that manufactures a flight environment indistinguishable from the real thing, but maintains insight. Imagine if simulators required pilots to take a drug that caused them to believe they were really flying. Pilots might be traumatised were they repeatedly faced with simulated 'emergency' situations; on the other hand, it might be dangerous were they to be in the real cockpit while doubting that fact! The ideal simulator must allow for a balance between resemblance and the ability to recognise what is really happening. What it creates should be believable without being believed, a kind of experiential verisimilitude.

We can learn something by comparing the aviation simulator to virtual soldiering, because something *different* is going on in the latter case. Aviation simulators aim at verisimilitude, but there is reason to doubt this is true in the military example. Since the whole point of the technology is to remove the soldier from harm's way, we might think that this includes the psychological harm of thinking one is on the battlefield. If that is right, there is a motivation to downgrade the quality of the simulation of the experience of battle. There is motivation to keep the soldier remote from the action, and his/her experience of it. Unfortunately this renders him/her less autonomous and more a tool of the state in its wartime pursuits. In this case the ideal virtual soldier would seem to be one who has both insight and protection from those experiences that have moral impact. But as I have argued, diminishing this moral impact compromises moral autonomy. Indeed, in a similar way to memory-deleting drugs, such enhancements also seem to hide the 'evidence' of moral experience by preventing them from forming, but in a different way: instead of making the memory remote from the person, in this case the person is made remote from the traumatic events. Yet the effect on moral autonomy is the same.

From a moral perspective there is in these cases a balancing act as well. Keeping a soldier away from the battlefield is obviously prudent. However, removing them to the point where they are rendered unable to decide and act as moral agents is to destroy their moral autonomy. Somewhere along this continuum lies a sweet spot between acting prudently and acting responsibly. That is the challenge for this case, and, I submit, the challenge for bio-technological soldiering in general.

References

Aquinas, Thomas. 1948. *Summa Theologica*. New York: Benziger Bros.
Augustine, St. 1958. *City of God*. New York: Doubleday.

Bublitz, Jan Christoph and Merkel, Reinhard. 2009. 'Autonomy and Authenticity of Enhanced Personality Traits', *Bioethics*, 23(6), 360–74.

DARPA 2012. *Department of Defense Fiscal Year (FY) 2013 President's Budget Submissions*. Justification Book Vol. 1.

Glover, Jonathan. 1988. *I: The Philosophy and Psychology of Personal Identity*. London: Penguin.

Kant, Immanuel. 1996. 'An Answer to the Question: What is Enlightenment? (1784)', in Gregor, Mary J. (ed.), *Practical Philosophy: the Cambridge Edition of the Works of Immanuel Kant*. Cambridge: CUP, 11–23.

Kennett, J. and Matthews, S. 2003. 'The Unity and Disunity of Agency', *Philosophy, Psychiatry & Psychology*, 10(4), 302–12.

Korsgaard, Christine. 2009. *Self Constitution: Agency, Identity and Integrity*. Oxford: Oxford University Press.

Lifton, Robert Jay. 1986. *The Nazi Doctors: Medical Killing and the Psychology of Genocide*. New York: Basic Books.

Lin, Patrick, Mehlman, Maxwell J. and Abney, Keith. 2013. 'Enhanced Warfighters: Risk, Ethics, and Policy'. Report prepared for the Greenwall Foundation, New York.

Walzer, M. 2004. *Arguing About War*. New Haven, CT: Yale University Press.

Williams, Bernard. 1981. 'Persons, Character, and Morality', in B. Williams, *Moral Luck*. Cambridge: Cambridge University Press.

Chapter 10

Ethical Considerations in
Military Surgical Innovation

Katrina Hutchison and Wendy Rogers

Surgical innovation raises a number of ethical challenges. There is the risk of harm to patients when a new procedure goes wrong or has unexpected consequences. There are challenges associated with informed consent, especially as the risk profile of new procedures may be difficult to assess. There is also the potential for conflicts of interest, particularly when the surgeon has a stake in the innovation. These ethical challenges take on new dimensions in the military context. The sort of surgery that occurs in military settings, especially combat zones, can differ from civilian surgery. Thus the sorts of innovations surgeons are doing – and the potential harms and risks to their patients – may follow a different pattern. Voluntary informed consent is particularly challenging, because of the intensely hierarchical structure of the military. Military surgeons face, within their role, conflicts of interest that do not confront civilian surgeons, because their duties to individual soldier patients may not align with their duties to the fighting force.

In the first section of this chapter we define surgical innovation. In the second section we provide a selected historical overview of surgical innovation in military contexts, in which we primarily focus on innovations in the care of wounded limbs and faces, looking at wartime developments in amputation, hand surgery and facial reconstruction surgery. In the third section of the chapter we discuss the ethical issues raised by surgical innovation in military contexts. We explore how issues such as harm to patients, informed consent and conflicts of interest (Johnson and Rogers 2012) manifest when surgical innovation occurs in the military context. We consider these issues with a view to future as well as past and present military surgical innovation. Our discussion of future issues focuses primarily on the ethics of enhancement, given that this is an area of current military research and raises important ethical issues. We make a number of concrete suggestions as to how the unique ethical challenges raised by surgical innovation in the military can be met.

What is Surgical Innovation?

Surgical innovation is a difficult concept to define for several reasons. First, surgery is multifaceted, and innovation can be associated with any of the different aspects of a procedure, including surgical techniques, surgical tools, implantable devices, technology, composition of the surgical team and operating conditions. Innovations in supportive activities such as anaesthesia, blood transfusion, imaging and wound care are also relevant to surgical outcomes. Second, it can be difficult to identify the point at which routine variation gives way to surgical innovation. It is a truism that the surgeon never does the same operation twice: procedures vary from patient to patient due to factors such as anatomical differences and co-morbidities, as well as surgeons' preferences for particular techniques or instruments. Routine variations of this kind can be difficult to distinguish from 'real' innovations – those more significant, systematic changes to techniques or tools that potentially alter the safety or effectiveness of the procedure. It can be difficult to predict which changes will impact upon the risk profile of a treatment – even the use of an existing tool for a new patient group or indication can significantly change the safety profile of the procedure: in a case in the UK, for example, a six-year-old child died after her routine splenectomy was performed using a morcellator. This tool had not previously been used either in children or for removal of the spleen, thus its use represented a significant innovation. A definition of surgical innovation should be able to capture these cases as well as more obvious innovations, such as the introduction of robotic surgery or the first hand transplant.

In response to these challenges we have elsewhere suggested drawing on Wittgenstein's notion of 'family resemblance' to develop an appropriately flexible definition of surgical innovation (Hutchison et al., *manuscript*). Like different games (Wittgenstein, 2009), different forms of surgical innovation are relevantly similar to one another but do not share an essential feature in common, thus any definition must recognise the different forms of surgical innovation. Our definition applies to new techniques, new tools and new devices. It also distinguishes several forms of newness that might obtain in each case – altogether new, new to anatomical location, new to patient group, new in the hands of an experienced surgeon and new to hospital. The first three can be considered 'real' innovation, whereas the last two can be thought of as personal or local forms of innovation. In civilian contexts the local forms of innovation are interesting because they raise a similar set of ethical and regulatory challenges as 'real' innovation – particularly issues associated with the learning curve for the surgeon and her team. In the context of this chapter, we focus on 'real' innovation, but note that local or personal innovation also occurs in the military, as when decisions are made about which procedures will be done by which surgeons and at which centres.

In addition to the variety of forms of surgical innovation, it is worth noting that a further complication arises insofar as innovations can be planned, serendipitous or heroic. Planned innovations are potentially recognisable before they occur, as the decision to innovate is made prior to the surgery. Serendipitous innovations are usually unplanned solutions to a problem that is discovered unexpectedly during patient care. An example of a serendipitous innovation is the invention of haemorrhage occluder pins – small, sterile titanium pins used to control sacral bleeding. These provide a highly effective solution to a life-threatening complication, and were developed after successful serendipitous use of sterilised thumb-tacks in China (Wang et al. 2009). Heroic innovations occur 'on the run' in response either to unexpected complications that occur during surgery or to the unique and critical needs of a trauma patient. Much surgical innovation in military contexts is likely to be heroic, given the prevalence of life-threatening wounds from gunshot and explosive devices. Many of the ethical questions associated with surgical innovation do not arise in cases of serendipitous or heroic innovation, as informed consent is often not possible in these cases, and the issue of patient harm is muted if the patient's life is at stake from injuries that will not respond to routine treatment.

Historical Overview of Surgical Innovation in the Military

Military contexts are associated with specific patterns of injury requiring surgical intervention, such as those associated with gunshot wounds and explosives including landmines, grenades and, increasingly in recent wars, improvised explosive devices (IEDs) (Manring et al. 2009). Due to the rarity (or non-existence) of some of these injuries outside of conflict situations, it is inevitable that innovation in their treatment will develop within military contexts. The high number of operations performed during times of conflict has also driven the development and diffusion of crucial medical services that support surgery, including anaesthesia, blood transfusion and wound care. In what follows we focus on three areas of surgical care that have seen significant innovation during wartime: amputation, hand surgery and facial reconstructive surgery. These examples do not provide an exhaustive picture of military surgical innovation; rather they provide the background for the discussion of ethical issues that follows.

Life-saving Innovation: Early Amputation in the Napoleonic Wars

During the Napoleonic wars, French surgeon Baron Dominique Jean Larrey developed triage, and was responsible for innovations in patient transport. In general his approaches focused on the wellbeing of the patient rather than returning soldiers to the battlefield, as his triage system treated the most badly wounded first irrespective of their prospects. His innovations focused on rapid access to care via well-equipped 'flying ambulances' (horse drawn) carrying medical personnel to provide immediate assistance. Larrey practiced the early amputation of wounded limbs to avoid tetanus and other infections. This was in contrast with the prevailing view that amputations should be delayed until infection had set in (Williams 1843, p. 226). Larrey's belief in early amputation was formed as a result of observing outcomes from wounds during his training (Welling, Burris and Rich 2010). This approach is typical of early innovative approaches to limb wounds insofar as it primarily aimed to save the life of the soldier rather than the function of the limb. Larrey was accused by contemporaries such as American surgeon James Mann of using amputation too widely, and using it too early for some indications

(Lancet 1938). Nevertheless it is widely accepted that his approach saved lives in comparison with the conservative approaches it replaced (Manring et al. 2009; Welling, Burris and Rich 2010). Although many of the procedures carried out at Larrey's field hospitals during the Napoleonic wars were undertaken in the context of emergency care, the introduction of the early amputation protocol can be seen as a planned innovation rather than heroic or serendipitous, as it was prospectively applied in response to a known problem.

Since Napoleonic times, the indications for amputation have become more restricted. This reflects innovations in vascular surgery, fracture treatment stabilisation (including surgical and non-surgical techniques), wound care and combating infection (Manring et al. 2009). However, the use of IEDs in wars in Afghanistan and Iraq, with associated complex and contaminated injuries, has contributed to a new rise in amputations. In the final section of the chapter, we argue that improvements in prosthetics might tilt the scales back in favour of amputation in some cases, especially where it is possible that a soldier with a prosthetic limb will be better, and more quickly, able to return to active duty than one subject to attempts at limb salvage.

Function-restoring Innovations: Hand Surgery in World War II

Hand surgery as a surgical specialty can be traced back to the influence of Norman T Kirk, surgeon general to the US Army in World War I. Kirk's experience as a doctor in World War I had led him to realise that treating hand injuries requires a combination of orthopaedic, plastic, vascular and neurological skills. He therefore recruited Dr Sterling Bunnell, a civilian surgeon with a special interest in hand surgery, to establish a US army program for treating for hand injuries. Nine specialist hand centres were set up by Bunnell in US army hospitals (Yakobina et al. 2008; Omer 2000; Newmeyer 2003), leading to a cohort of expert surgeons who later became founding members, and many of them presidents, of the American Society for Surgery of the Hand.

Advances in hand surgery reflect more general developments in military surgery. Prior to World War II, the primary focus on saving life meant that a uniform protocol was used for treatment of limb wounds. However, once viable treatments for life-threatening infections and blood loss existed, surgeons were able to shift their focus to more complex procedures aimed at restoring function. As described above, during World War II there was increased focus on restoration of hand function for soldiers with hand injuries. Today the focus of military surgery and prosthetics research for hand loss remains on restoration of function, but the goal of enhancement looms as a possibility in the near future. The United States Defense Advanced Research Projects Agency (DARPA) has a current project called 'revolutionizing prosthetics'. At this stage the aim of the project is to restore full functionality to those who have lost a hand, through the use of complex prosthetic devices. The project has already yielded one FDA-approved, futuristic-looking prosthetic hand (Guizzo 2014). It seems possible that future prosthetics projects may aim at enhanced function in some domains of activity. Domain-specific functional enhancement has arguably already been achieved with prosthetic legs, as the controversy associated with Oscar Pistorious' bid to run in the Olympics on his carbon blade prosthetics revealed (Burkett, McNamee and Potthast 2011). In the military context, some soldiers in the USA have already returned to active duty with prosthetics, demonstrating that an adequate level of function has been achieved for some tasks (Manring et al. 2009). Even if current prosthetics do not lead to improved performance, they might nonetheless offer an advantage in terms of resistance to some of the ills of injury. Although a prosthetic device can be damaged by a bullet or an IED, such damage does not cause the pain, blood loss or risk of mortality to the soldier suffered by similar injuries to the flesh. Furthermore, simple 'enhancements' such as a torch or global positioning system, could presumably be built into current day prosthetic hands.

From Functional to Cosmetic Enhancement: Facial Reconstruction and Plastic Surgery in Military and Civilian Contexts

Reconstructive plastic surgery developed as a discipline largely in response to the high number of maxillofacial injuries sustained by soldiers in trenches during World War I (Mazzola and Kon 2010).[1] Leaders of innovation in facial reconstruction during the war included French, Dutch, German, Russian and Italian surgeons. French surgeon Hippolyte Morestin pioneered interdisciplinary teamwork between general surgeons and those with special expertise in dental and oral surgery. His partnership with oral surgeon Auguste Valadier at the Val de Grace military hospital in Paris inspired the leading figure of plastic surgery, New Zealander Harold Gillies (Mazzola and Kon 2010, p. 889). Gillies became committed to the specialist treatment of facial injuries. He lobbied for separate wards at the Cambridge Military Hospital in Aldershot, and when the war office refused to issue special facial injury casualty tags, he created them himself, addressing them to his wards, and instructing the war office as to which casualties to send to him (Bamji 2006). He later set up a hospital for treating maxillofacial injuries, which became a referral centre for these injuries across Europe. Like the French doctors who inspired him, his team was multidisciplinary and included dental surgeons, radiologists and highly qualified anaesthesiologists, as well as artists, sculptors and photographers. The latter were involved in planning reconstructions and documenting results (Bamji 2006). Gillies and his colleagues developed and refined reconstruction techniques, but it was not only the plastic surgeons at the facility who innovated. Facial injuries required inventive approaches to anaesthesia as standard techniques for delivering chloroform or ether used a mask that covered the face, making surgery impossible. In response, anaesthesiologist Ivan Magill

---

    1   Prior to the war there were cosmetic plastic surgeons but their work was disreputable and they did not have an association or other markers of a distinct surgical specialty (Mazzola and Kon 2010).

developed the new technique of endotracheal intubation (Mazzola and Kon 2010). Thus wartime treatment of facial injuries laid the foundations for the now standard use of intubation to deliver anaesthetic gases and oxygen during surgery.

During the war, some centres focused on restoring function whereas others (including Gillies') addressed both function and aesthetics. The sort of facial reconstructions undertaken in response to war injuries are not common in civilian settings. However, the skills involved in reconstructing delicate facial features such as cheeks, noses, mouths, ears and eyelids are applicable to civilian cosmetic facial surgery. The relationship between innovations in plastic surgery during war, and civilian use of these techniques is interesting, and raises ethical questions about how the burden of innovation is distributed across a population, especially when innovations developed in response to the unique characteristics of war wounds find unforeseen application in civilian surgical enhancement.

We turn now to examine the ethical issues arising from military surgical innovation.

Ethical Issues

Surgical innovation is largely motivated by the desire to improve upon existing surgical interventions. Improvements can occur across various dimensions of surgery. Common targets of surgical innovation include: decreasing the duration of surgery; using less invasive techniques; shortening hospital stays; improving patient outcomes; or curing the previously incurable. As the examples discussed in the previous section suggest, the targets of surgical innovation have changed as medical science has developed: Larrey's amputation protocol was aimed at controlling life-threatening infection, whereas the advent of antibiotics has rendered the surgical control of infection less urgent. However, the primary ethical motivation for surgical innovation has not changed. Beneficence – that is, better surgical experiences and outcomes for individual patients – has always been the common primary goal of innovating surgeons. Of course, there are also other potential or actual motives such as the surgeon's desire to pioneer new techniques and to be recognised as a leader in the field; the desire (individual or corporate) to develop a financially rewarding innovation; and the need to find new efficiencies in the use of healthcare resources. At times such motives can lead to conflicts of interest, as we discuss below. In the context of military surgical innovation, an additional consideration arises. This relates to the patient qua soldier, and the need of the military for soldiers who are as fit as possible or who can return to active service as rapidly as possible following injury. Thus, as well as the recognised ethical issues arising from civilian surgical innovation (Johnson and Rogers 2012), military surgical innovation introduces the ethical complexity of interventions performed on patients in order to benefit not only themselves, but also the military those patients serve: there is an imperative to combine good medicine with good tactics (Blackbourne et al. 2012, p. S389). Our discussion here focuses on issues of harms to patients/soldiers, informed consent for surgical innovation in the military context, and conflicts of interest associated with military surgical innovation.

Before turning to specific ethical issues, it is worth considering whether or not there is a duty to innovate in surgery, and if so, how the military context might affect this. Unlike at least some branches of medicine, surgeons are both diagnosticians and intimately associated with the quality of the intervention that they deliver; they are part of the therapeutic modality and as such, personally responsible for the quality of the surgery they perform. Whether or not the patient does well is dependent to a significant degree upon the skills of the individual surgeon. This feature is evident in the historical accounts of military surgical innovation discussed in the previous section: Larrey and Gillies were not only innovators in terms of changing care protocols and extending the treatment options available, they were also regarded as extremely talented surgeons (Williams 1843; Bamji 2006). This feature of surgery, that surgeons themselves are an inherent part of the therapy, creates an obligation for surgeons constantly to strive to improve their techniques, through research, innovation and repetition. But this can be ethically challenging for the following reasons. First, the outcomes of surgical innovations cannot always be predicted in advance, making it difficult for surgeons to be confident that they are acting in the best interests of their patients when they innovate. Second, adopting new techniques entails a learning curve during which the first cohort of patients exposed to the new technique will be disadvantaged compared with patients treated when the surgeon is more practiced. Third, as the boundaries between innovation and research are not well demarcated, this can lead to role confusion for the surgeon who is both researcher and practitioner (Rogers and Johnson 2013). In the military context, the pressures to innovate may be magnified. Combat in war may provide the only source of particular types of injuries and wounds, and thus the only opportunity to develop new treatments. As mentioned above, the use of IEDs in Afghanistan and Iraq has led to new types of injuries with specific characteristics including limb amputations and extensive abdominal and genitourinary soft tissue injuries contaminated with dirt and other debris (Blackbourne et al. 2012). Given the unique characteristics of these injuries, any innovations in managing soldiers wounded by IEDs has to occur in the care of those patients; it is not possible to generate the relevant knowledge from another patient cohort as there is no comparable civilian weapon or pattern of injuries. Thus in situations of military conflict that generate unique injuries, the duty to innovate falls heavily upon military surgeons, as it is only in this context that relevant innovation can occur. We now turn to specific ethical issues arising through surgical innovation.

Patient Harms and Benefits

Innovations in surgery have the potential to benefit patients, but equally, innovations may lead to patient harm. Patients may have increased morbidity and mortality following innovative procedures, despite the beneficent intent of the innovation. The history of surgical innovation is replete with examples of well-intentioned innovations that left patients either no better or worse off than if they had received the standard procedure. Examples include ligation of the internal mammary artery for angina which was eventually found to be no better than placebo (Beecher 1961), freezing of the stomach lining as a treatment for gastric ulcers (Frader and Caniano 1998) and insertion of the DePuy articular surface replacement hip prosthesis, which had high rates of failure and other complications (Cohen 2012). The underlying challenge regarding innovation is that the effects of the innovation are unknown at the time that it is performed. Thus it is impossible to predict whether or not the innovation will be better than the existing alternative, or whether patients will be harmed. For innovations that occur in emergency life-threatening situations, trying something new can be justified as there may be no alternative. In these situations, even if the innovation is unsuccessful, the patient is no worse off than they would have been absent the innovation. Here the harm to the patient is not the direct result of the innovation; rather the innovation failed to ameliorate a life-threatening emergency. Planned innovations bear a different burden of responsibility, in that with a planned innovation, there is time to consider and anticipate potential harms, even though these may not always be accurately predicted. However, it is not possible to identify in advance unanticipated harms, thus even with the best planned innovation there will always be the risk of harm. Where the innovation is performed to enhance rather than to address an existing health problem, the responsibility becomes correspondingly higher: surgical enhancement involves a healthy patient (in the case of soldiers, usually an extremely fit person) subjected to a surgical intervention that is not therapeutically indicated. Such intervention entails at least some morbidity (pain, scarring, exposure to anaesthesia, anatomical modification), not all of which is reversible. Thus the provision of innovative surgical enhancements requires critical and independent assessment of the potential for harms, the justifications for the enhancement and the likely effects (physical, psychological) on the recipient. Indeed, it is unclear whether surgical innovation that aims at enhancement can be thought of as motivated by beneficence, which we identified as the primary goal of past and present surgical innovation.

As well as unavoidable harms from the procedure itself, innovative surgery can be harmful to patients during its development phase due to the learning curve. When surgeons take up a new procedure, they are more likely to encounter complications and take longer, than once they have fully mastered the new technique. Thus even innovations that prove in time to be safer and more effective than their predecessors can lead to harm for patients treated in the early phases of the introduction of the innovation. Laparoscopic cholecystectomy provides a clear example of this; when the procedure was first introduced there were high rates of bile duct damage compared to the open procedure (Moore, Bennett and Meyers 1995). Once established, however, laparoscopic cholecystectomy rapidly achieved better outcomes for patients than the open procedure. Estimates of the number of cases needed to perfect surgical techniques vary, but may be as high as 200 (Voitk, Tsao and Ignatius 2001).

The learning curve presents an additional ethical challenge when the group of patients who undergo surgery during the learning curve differs demographically from groups who will later benefit from the surgery. The development of facial reconstructive surgery presents just such a case, as contemporary recipients of cosmetic plastic surgery, that is, healthy individuals whose aim is to improve their appearance, benefit from techniques developed on wounded soldiers. It is not obvious how to address this ethical challenge. There does not seem to be anything inappropriate about the early facial reconstructive surgery performed on the soldiers, undertaken in the interests of securing the best possible functional and aesthetic outcomes for those patients. Nor do current cosmetic procedures harm or disrespect past soldiers in any meaningful sense. Given the high number of operations performed in times of war, perhaps the first step is to recognise that soldiers have been and will continue to be over-represented in the learning curves for some surgical procedures, particularly procedures that aim to repair or reconstruct wounded bone and tissue. It is important to ensure that the lessons learned on soldiers – both those with civilian application and those unique to conflict injuries – stay learned, to minimise the unnecessary suffering of future soldiers. Military history is full of innovations that were forgotten and had to be relearned in the next war, including Larrey's innovations in triage, transport and en-route care, which were ignored by the British during the Crimean War (Manring et al. 2009). In terms of recognising the over-representation of soldiers in surgical learning curves for some widely used civilian procedures such as cosmetic surgery, most countries have social practices for recognising the sacrifices of soldiers, as well as social structures to support returned soldiers, and these might be the most appropriate forums for recognising this particular contribution.

The deleterious impact of the learning curve upon surgical outcomes also leads to challenges in knowing when to evaluate innovations and when to abandon them as harmful or ineffective. If no allowance is made for the learning curve, we risk abandoning procedures that may turn out to be beneficial once surgeons have perfected their techniques, but on the other hand, undue allowance for initial poor outcomes may lead to harmful innovations remaining in practice for prolonged periods. This challenge applies equally in military and civilian circumstances. One proposed solution to this is a system of data collection and evaluation starting when the innovation is first implemented (McCulloch et al. 2013). This approach is being used with some success in the US military context to evaluate a new system of managing combat casualties known as the Joint Trauma System (JTS). The JTS is described as 'a novel systematic and integrated approach to organize and coordinate combat casualty care' (Blackbourne et al. 2012, pS388). A Joint Theater Trauma Registry (JTTR), commenced in 2005, supports the JTS. The JTTR records a variety of data providing vital information to drive improvements in clinical care, as well as generate new knowledge (Blackbourne et al. 2012). Blackbourne et al. note that although the JTTR was not set up as a research data tool, knowledge generated from the registry about the JTS has revolutionised field trauma care and fostered advances in military medicine. The JTTR provides an excellent model for maximising knowledge about the introduction of an innovation. The highly regulated context of the military has enabled the development of a comprehensive registry tracking the outcomes of the new trauma management system, thus maximising benefits and leading to the earliest possible identification of harms.[2]

Informed Consent

Informed consent refers to the permission given by a patient for a medical or surgical intervention. Healthcare interventions are ethically justifiable only if the patient concerned is competent to make a decision; has been provided with, and fully understands relevant information about the procedure; and is making the decision freely and voluntarily. Military surgical innovation poses a number of challenges to informed consent. First, patients may not be competent to give consent for innovations provided as part of the management of trauma, as they may be unconscious, in pain or under the influence of analgesic drugs. In these situations, consent is not possible; the surgical innovation must be ethically justified on the grounds that it is in the patient's interests. The prospective collection of data on military surgical interventions, particularly innovative interventions, as recommended in the above section, can act as a further safeguard of the interests of soldiers receiving unplanned innovative care. In contrast, patients are usually competent to consent to planned innovations, such as procedures to implant rehabilitative or enhancing prostheses. However, there are other threats to informed consent in these situations, relating to the provision and understanding of information, and the voluntariness of the decision. As detailed above, by their very nature, there are unknown risks associated with surgical innovations. Even where innovations are meticulously planned in advance, it is impossible to eliminate the 'unknown unknowns' – unanticipated adverse outcomes. This feature leads to a moral obligation to consider whether or not innovative surgery is a type of research, and should therefore use the standards of research consent rather than those of routine clinical care (Lotz 2013). The standards for consent to research are more demanding than those of clinical practice, in part to ensure that participants understand that the research is aimed at generating knowledge about the intervention, and that, in advance of that process, it is not possible to provide a full account of the likely harms and benefits.

A second and serious threat to valid informed consent in the context of elective military surgical innovations relates to voluntariness: valid consent

---

2   There were also meticulous records of treatment of vascular injuries during the American Civil War. The Surgeon General had an academic background, and insisted upon the keeping of these records. The records relating to vascular injuries have been described as 'the largest experience with vascular injuries ever described' (Blaisdell 2005 p. S21).

must be freely given, absent of any coercion or undue influence. The military is intensely hierarchical and fosters a culture in which junior enlisted personnel must exhibit deference and obedience to their superiors who exert considerable power over all aspects of their life (Spence 2007). This creates a challenge in terms of consent. First, junior personnel may interpret the offering of a specific innovative treatment akin to an order – as something that they must do because their superiors have suggested it. This consideration may be less important in the case of innovations aimed at treating existing disorders where the desire to remedy an existing condition provides at least some justification for the intervention. However, even these cases present challenges: a soldier with an injured hand might find it extremely difficult to choose between retaining their own disabled hand that nonetheless has some normal features such as skin sensitivity, versus being fitted with a prosthetic that lacks these features but has greater functionality. In this situation, the recommendations of a military superior might be decisive. More importantly, the therapeutic justification does not apply in the context of innovations aimed at enhancement, as the solider is healthy in the absence of the intervention. An offer of an enhancing innovation may seem very much like an order. There may be implicit or explicit inducements (or threats) accompanying the offer, such as career advancement, more favourable deployments or better conditions. These may be real or imagined, but even if there are no specific material consequences of refusing an enhancing innovation, the soldier's belief that there will be consequences compromises the voluntariness of their consent. As well as the pressures imposed by the military hierarchy, there are also peer pressures that may compromise consent. Junior military personnel are often a very close-knit community; any refusal to participate in a program, such as enhancement, may lead to fear of being ostracised (Spence 2007). The effects of exclusion from a peer group may impact seriously, especially on personnel in isolated and hazardous situations who rely upon their peers for psychological and social support, as well as to risk their lives for each other.

For these reasons, it seems warranted to suggest that military surgical innovations, especially those aimed at enhancement rather than therapy, be held to the standards of research consent. The demarcation between research and innovation is indistinct at the best of times; in the military context where there are clear and credible threats to the voluntariness of consent, adhering to research standards will offer greater protection to military personnel than standard consent to clinical care. This will not address all of the ethical concerns related to consent, as military personnel are recognised as vulnerable research populations for the reasons outlined above, but it will help to ensure that measures are taken to mitigate the factors that may compromise informed consent.

Research in military contexts poses challenges, including challenges for informed consent and logistical challenges including resource limitations. Yet it is important that the demands of ethical oversight do not stifle innovation in these contexts. When Ambrose Pare ran out of 'seething oil' with which to treat war wounds in 1536, it is unlikely that he sought informed consent from the wounded soldiers treated with his innovative alternative: 'a "healing salve" made of yolk of eggs, oil of roses and turpentine' (Pruitt Jr 2006, p. 717). Yet this innovation transformed the care of war wounds, without the benefit of a formal research trial. Until recently, there was little scope for conducting prospective research on surgical innovations provided as part of surgical care in combat zones (Hatzfield et al. 2013). Now, with the introduction in 2005 of a human research protection plan (HRPP), prospective research on US soldiers in combat zones is possible. This ensures that advances such as Pare's are possible, while mitigating some of the risks. Research on humans in the US military falls under the care and control of the US Army Medical Research and Materiel Command, which provides ethical review and regulatory oversight for all research involving military personnel (Hatzfield et al. 2013). There are considerable logistical and other challenges in performing research in a combat zone. The research must not interfere with operational procedures, patient care or protection of the health of the military personnel, and must be carried out within the resource limitations of the combat zone. Approval of such research by ethical review boards raises a number of further issues, which we discuss in the next section. Nonetheless, as Hatzfield et al. show, it is possible to perform prospective research into the care of wounded soldiers within a research ethics framework.

Conflicts of Interest

Innovation in surgery creates the potential for conflicts of interest. Surgeons have a primary duty to act in the best interest of their patients. This duty can come into conflict with other duties related to their role, such as that of educating and training the next generation of surgeons, developing and improving surgical techniques through innovation or collecting data for quality control. These types of conflicts are known as within-role conflicts, or conflicts of obligation, because the competing interests arise from a legitimate part of the practitioner's role (Rogers and Johnson 2013; Lo and Field 2009). In contrast, external conflicts of interest occur when the competing interest is unrelated to the role of surgeon, such as financial gain, or family concerns. In general, conflicts of interest – both external and internal – are poorly addressed. Most policies require disclosure of conflicts, but this has been shown to be an inadequate solution (Johnson and Rogers 2014). Surgeons undertaking innovation in military contexts are not immune to these general challenges, but as mentioned above, have an additional role-related duty which may compete with that of patient care, and that is their duty to the military. As Annas puts it, it is not clear whether: 'physicians in the U.S. military are physicians first, soldiers first, or physician–soldiers' (2008, p. 1087). Historically this conflict is apparent in a manual on the goals of triage for British World War I surgeons, which instructed them to prioritise conservation of manpower first and the interests of the wounded second (Manring et al. 2009, p. 2170). Until recently, physicians within the US military have been considered bound by internationally recognised norms of medical ethics; that is, they were physicians first. It was just such norms that provided the grounds for indicting Nazi doctors (Annas 2007). Recently, the US military has distanced itself from international norms and issued instructions, for example on force feeding, which are incompatible with both US and international standards of medical ethics (Annas 2007). This leaves US military medical personnel in a potentially precarious situation in which they may be forced to choose between disobeying orders or violating the ethical foundations of their profession.

The development of prosthetics, such as the DARPA prosthetic hand discussed in the previous section, exemplifies how enhancement might put pressure on the relationship between what is in the best interest of the fighting force and what is in the best interest of the individual soldier. In the case of a prosthetic limb these interests can come apart even when the prosthetic serves a therapeutic purpose. Whereas a soldier with a serious hand injury might be happier with a disabled but real hand that retains skin sensitivity and other characteristics of human flesh, the interests of the military could differ. Even a fairly rudimentary prosthetic hand that does not enhance ordinary functionality might offer advantages from a military perspective. For example, the absence of skin sensitivity to heat, cold and pain could be an advantage, as could the fact that a gunshot or explosive 'injury' to the prosthesis is unlikely to pose an immediate risk to the life of the soldier. The wording of the aims of DARPA's current Revolutionizing Prosthetics program reveals an apparent ambiguity between the interests of the individual patient and the interests of the military: 'The Revolutionizing Prosthetics program is ongoing and aims to continue increasing functionality of the DARPA arm systems so service members with arm loss may one day have the option of choosing to return to duty' (Defense Advanced Research Projects Agency [DARPA] 2014). The emphasis on personal choice implies respect for the wellbeing of the individual, but in fact there is no mention of wellbeing. Use of dispassionate language, as in the phrase 'service members with arm loss', seems to underplay the humanity of injured soldiers, while the focus on the possibility that they might 'choose' to return to duty suggests underlying military interests.

Just how the within role conflicts for military surgeons might play out in the case of innovative surgical enhancements of military personnel is unclear. Surgeons who are strongly committed to the enhancement, perhaps because they have been involved in its development, who feel that enthusiasm is necessary to protect their military careers, or believe that the enhancement may be lucrative in civilian markets, will have a conflict of interest in the care of soldiers for whom the enhancement is recommended. At a minimum, soldiers offered militarily motivated surgical enhancements should have access to independent surgical advice. If the enhancements are classified, it may be difficult to find an independent surgeon, as those outside of the military will not have access to research data or other information about the innovation. In this case, it would be warranted for the military to ensure that there are 'enhancement sceptics' within the ranks of its medical corps, who are cleared for access to relevant information but not involved in the development of the enhancements. However, as any surgeon within the military is subject to the kinds of pressures documented earlier, including peer pressure and fear of retribution, this proposed solution is far from ideal.

A further potential conflict of interest arises regarding ethical review of research involving military surgical innovations. Hatzfield et al. (2013) note that the US Army Medical Research and Materiel Command (MRMC) provides ethical review for research in the military. We are unaware of the exact make up of the ethical review board of the MRMC, but if this is composed entirely of individuals from within the military, there may be a conflict of interest between military objectives and the protection of research participants. One of the roles of research ethics review boards is to provide independent assessment of the potential harms to participants, and to weigh these up against the potential gains to knowledge from the research. The requirement of independence will not be met if all members of the review board are military personnel. More importantly, there is a danger that the balance of review board decisions will be skewed in favour of military interests.

Conclusion

In this chapter we have argued that surgical innovation in the military context gives rise to unique versions of some of the ethical challenges raised by surgical innovation in civilian settings. The rarity (or indeed uniqueness) in civilian settings of the types of injuries that occur in conflicts means that military contexts are the only places in which innovations for treating these injuries can occur. Thus soldiers are disproportionately represented in the learning curves for innovations in the treatment of traumatic injuries. This poses risks of harm to soldiers, as well as potential inequity insofar as members of the wider population later benefit in unforeseen ways from the techniques developed on the battlefield, as has occurred in the case of cosmetic surgery. In light of these ethical challenges, it is desirable that we maximise the information we harness from military surgical contexts, and that the lessons of military surgery remain learned to avoid unnecessary harm to future soldiers. We thus recommend comprehensive prospective databases for recording military surgical interventions, particularly those that are innovative.

Military contexts pose unique challenges for informed consent, given the intensely hierarchical structure of the military, and the disproportional representation of those of lower rank as patients. In light of this, we recommend that the standard of consent for planned surgical innovation in military contexts should always be the same as the standard for research. Research is held to higher standards of consent than other healthcare interventions, making it the best option available for ensuring that consent is voluntary and informed.

Another ethical issue we have raised is that of within-role conflicts of interest. These are an issue for surgical innovation in general, and have a unique dimension in military contexts. Military surgeons are subject to within-role conflicts between their duty to the patient and their duty to the military. In cases where these dual duties pull apart – for example where enhancement is in the best interest of the military but perhaps not the patient – there is a risk that the patient's best interest will not be paramount. This risk may be compounded insofar as military surgeons themselves are influenced by the hierarchical military structures within which they practice. We thus recommend that in addition to research standard consent procedures, it is also important for patients to be provided with an independent second opinion regarding any proposed innovative treatment. This will be particularly important in cases of surgical enhancement, which lack therapeutic justification. Finally, we note that research ethics review within the military may also raise conflicts of interest where such review is provided exclusively or predominantly by military personnel. A strategy for addressing this issue is to include non-military members on ethical review boards for military surgical research.

References

Bamji, A. 2006, 'Sir Harold Gillies: Surgical pioneer', *Trauma*, 8, 143–56.

Beecher, H.K. 1961, 'Surgery as placebo: A quantitative study of bias', *Journal of the American Medical Association*, 176: 1102–7.

Blackbourne, L.H., Baer, D.G., Eastridge, B.J., Butler, F.K., Wenke, J.C., Hale, R.G., Kotwal, R.S., Brosch, L.R., Bebarta, V.S., Knudson, M.M., Ficke, J.R., Jenkins, D. and Holcomb, J.B. 2012, 'Military medical revolution: Military trauma system', *Journal of Trauma and Acute Care Surgery*, 73(6) SUPPL. 5: S388–S394.

Blaisdell, F.W. 2005, 'Civil War Vascular Injuries', *World Journal of Surgery*, 29, s21–s24.

Burkett, B., McNamee, M. and Potthast, W. 2011, 'Shifting boundaries in sports technology and disability: Equal rights or unfair advantage in the case of Oscar Pistorious?', *Disability and Society*, 26(5): 643–54.

Cohen, D. 2012, 'How safe are metal-on-metal hip implants?' *BMJ*, 344: 1–5.

Defense Advanced Research Projects Agency [DARPA] 2014, *Revolutionizing Prosthetics*, DARPA, Virginia, USA. http://www.darpa.mil/Our_Work/BTO/Programs/Revolutionizing_Prosthetics.aspx

Frader, J.E. and Caniano, D.A. 1998, 'Research and innovation in surgery', in L.B. McCullough, J.W. Jones and B.A. Brody (eds), *Surgical Ethics*. New York: Oxford University Press, 217–41.

Guizzo, E. 2014, 'Dean Kamen's "Luke Arm" Prosthesis Receives FDA Approval', *IEEE Spectrum Online*, web log post, 13 Mai, viewed 10 July 2014, http://spectrum.ieee.org/automaton/biomedical/bionics/dean-kamen-luke-arm-prosthesis-receives-fda-approval

Hatzfeld, J.J., Childs, J.D., Dempsey, M.P., Chapman, G.D., Dalle Lucca, J.J., Brininger, T., Tamminga, C., Richardson, R.T., Alexander, S. and Chung,

K.K. 2013, 'Evolution of biomedical research during combat operations', *Journal of Trauma and Acute Care Surgery*, 75(2) SUPPL. 2, S115–S119.

Hutchison K., Rogers, W., Eyers, A. and Lotz, M. 'Getting clearer about surgical innovation: A new definition and a new tool to support responsible practice', *Annals of Surgery*, Online Early, 25 Feb 2015 (doi: 10.1097/SLA.0000000000001174).

Johnson, J. and Rogers, W. 2012, 'Innovative Surgery: The ethical challenges', *Journal of Medical Ethics*, 38: 9–12.

Johnson, J. and Rogers, W. 2014, 'Joint issues – conflicts of interest and the ASR hip', *BMC Medical Ethics* (accepted 17th July 2014).

Lancet, 1938, 'Feats of amputation', *The Lancet*, 232(5992): 31.

Lo, B. and Field, M.J. 2009, 'Committee on Conflict of Interest in Medical Research, Education, and Practice: Institute of Medicine', in B. Lo and M.J. Field (eds), *Conflict of Interest in Medical Research, Education and Practice*. Washington DC: The National Academies Press.

Lotz, M. 2013, 'Surgical innovation as sui generis surgical research', *Theoretical Medicine and Bioethics*, 34(6): 447–59.

Manring, M.M., Hawk, A., Calhoun, J.H. and Andersen, R.C. 2009, 'Treatment of War Wounds', *Clinical Orthopaedics and Related Research*, 467: 2168–91.

Mazzola, R.F. and Kon, M. 2010, 'EURAPS at 20 years. A brief history of European Plastic Surgery from the Societe Europeenne de Chirurgie Structive to the European Association of Plastic Surgeons (EURAPS)', *Journal of Plastic, Reconstructive and Aesthetic Surgery*, 63: 888–95.

McCulloch, P., Cook, J.A., Altman, D.G., Heneghan, C., Diener, M.K., on behalf of the IDEAL Group 2013, 'IDEAL framework for surgical innovation 1: the idea and development stages', *BMJ* (Clinical research ed.), 346, f3012.

Moore, M,J, Bennett, C,L, and Meyers, W,C, 1995, 'The learning curve for laparoscopic cholecystectomy', *American Journal of Surgery*, 170(1): 55–9.

Newmeyer, W.L. 3rd, 2003, 'Sterling Bunnell, MD: the founding father. *Journal of Hand Surgery*', 28(1): 161–4.

Omer, G.E. 2000, 'Development of hand surgery: Education of hand surgeons', *Journal of Hand Surgery*, 25(4): 616–28.

Pruitt Jr, B.A. 2006, 'Combat casualty care and surgical progress', *Annals of Surgery*, 243(6): 715–29.

Rogers, W. and Johnson, J. 2013, 'Addressing within-role conflicts of interest in surgery', *Journal of Bioethical Inquiry*, 10(2): 219–25.

Spence, D.L. 2007, 'Ensuring respect for persons when recruiting junior enlisted personnel for research, *Military Medicine*, 172(3): 250–53.

Voitk, A.J., Tsao, S.G.S. and Ignatius, S. 2001, 'The tail of the learning curve for laparoscopic cholecystectomy', *American Journal of Surgery*, 182(3): 250–53.

Wang, Q., Shi, W., Zhao, Y., Zhou, W. and He, Z. 1985, 'New Concepts in Severe Presacral Hemorrhage During Protectomy', *Archives of Surgery*, 120(9), 1013–20.

Welling, D.R., Burris, D.G. and Rich, N.M. 2010, 'The influence of Dominique Jean Larrey on the art and science of amputations', *Journal of Vascular Surgery*, 52: 790–93.

Williams, R. 1843, 'Life and works of Baron Larrey', *Lancet*, 40: 224–9.

Yakobina, S.C., Yakobina, S.R., Harrison-Weaver, S. 2008, 'War, What is it Good for? Historical Contribution of the Military and War to Occupational Therapy and Hand Therapy', *Journal of Hand Therapy*, 21(2): 106–13.

PART IV
Law, Responsibility and Governance

Chapter 11
Assigning Responsibility in Enhanced Warfare

Alex Leveringhaus

In recent years, researchers in neuroscience and related disciplines have gained a better understanding of the human brain and the neurophysiological processes involved in human decision-making (Taylor 2012). These scientific developments have been accompanied by a bioethical debate on the moral permissibility of human enhancement (Agar 2014; Buchanan 2011; Harris 2007; Persson and Savulsecu 2012). Strikingly, this debate has largely focused on enhancement in civilian contexts, yet the military is likely to have a strong interest in the development of enhancement techniques (Moreno 2012; Blank, 2013, pp. 218–27). After all, militaries throughout the ages have tried to enhance the fighting capacities of their personnel (Grossman 2009). Whether through strenuous physical training, or the administration of methamphetamine-based 'Go Pills' to US fighter pilots (Friscolanti 2005), militaries across the world have tried to make their soldiers run faster, jump further, fight harder and stay focused for longer. Fighting in war consists in creating asymmetries between one's own troops and enemy troops, thereby forcing the enemy into submission. In the context of the above, why shouldn't the military use what I refer to as 'military human enhancement' (MHE) in order to create advantages on the battlefield?

The answer to this question cuts across a number of complex theoretical issues. In this chapter, I look at one of these issues, the assignment of responsibility in enhanced warfare. In general, there are two ways in which responsibility can be assigned. When one assigns responsibility in a *backward-looking* sense to an agent, one either credits that agent for an event or blames him for that event (retrospective responsibility). When one assigns responsibility in a *forward-looking* sense to an agent, one makes the agent responsible for the performance of a particular task or the wellbeing of another agent in the future (prospective responsibility). Now, the problem with some emerging military technologies, and not just MHE, is that they might make it difficult to assign responsibility in war (Sparrow 2007). This may result in so-called 'responsibility gaps', where no one is responsible for the use of force. Hence one question this chapter seeks to tackle is whether MHE gives rise to responsibility gaps.

The chapter proceeds as follows. In the first part, I outline a number of background considerations on the nature of responsibility, in general, and just war theory, in particular. In the second part, I examine the implications of MHE for the moral agency of enhanced soldiers. Here I largely focus on retrospective responsibility. In the third part, I discuss the assignment of prospective responsibility.

Background Considerations

In this part of the chapter, I outline four background considerations that are crucial for my subsequent analysis. The first pertains to the relationship between three senses of responsibility, *causal*, *moral* and *legal*. Causal responsibility means that an agent is causally implicated in an event. However, causal responsibility for an event is neither necessary nor sufficient to be morally or legally responsible for that event. Agents might have agreed in advance that a particular person is morally or legally responsible for x, even if another person causes x. In this case, causal involvement in x is not necessary in order to be legally and morally responsible for x. Further, an agent may bring about an event without exercising agency. For instance, a person whose body becomes a human missile due to a strong gust of wind and squashes another person to death is causally but not necessarily morally and legally responsibility for the loss of life (McMahan 1994; Otsuka 1994): it was a freak accident. In this case, causal involvement is not sufficient to be legally or morally responsible for the death.

In war, assignments of causal responsibility are not always easy. For instance, in an exchange of fire with enemy troops, it is often not clear whose bullet killed the enemy. Enhanced warfare might exacerbate this problem because soldiers might fight with greater speed, over longer distances, or operate very complex combat technology.[1] Fortunately, there might be ways to strengthen the ability to assign causal responsibility. The use of helmet cameras leads the way here. The same goes for 'black boxes' installed in complex combat systems. However, it is important to be aware that, even if it is sometimes impossible to assign causal responsibility in enhanced war, it does not automatically follow that moral and legal responsibility are equally unassignable. It is thus possible that a soldier is morally or legally responsible for, say, killing an enemy combatant, even if it is not clear who pulled the trigger.

Compared to causal responsibility, the relationship between moral and legal responsibility is more difficult to ascertain. As a rule of thumb, the distinctive feature of legal responsibility is that it involves an element of institutional authority. Only select institutions are justified in enforcing the law by assigning legal responsibility to agents. Punishment for war crimes, for instance, is the task of an authorised legal body, not private individuals. By contrast, assignments of moral responsibility, with which this chapter is concerned, are open to all members

---

1   Swarm technology, where a single operator controls a number of machines that move in a synchronised fashion, serves as a useful example. This type of technology may push human cognitive ability to the limit. MHE might be one way to enable soldiers to operate ever more complex machinery.

of a community, and, for cosmopolitans in particular, potentially to all citizens of the world. This is important because it makes it possible to criticise the armed services and the government under whose command they fight, notwithstanding a lack of legal authority.

The second background consideration concerns the moral standard against which assignments of moral responsibility in enhanced warfare are made. Rejecting pacifism, I approach enhanced warfare from a just war perspective. Just war theory operates with two main normative frameworks that regulate the use of force. *Jus ad bellum* governs the declaration of war, while *jus in bello* governs the conduct of war. The topic of this chapter – assigning responsibility *in* enhanced warfare – means that I focus on *jus in bello*. New military technologies always raise the question whether compliance with the key principles of *jus in bello* – *discrimination* (the obligation to distinguish between legitimate and illegitimate targets), *proportionality of means* (the obligation not to cause excessive damage) and *necessity* (the obligation to abstain from causing militarily unnecessary harm) – is enhanced or threatened. MHE is no different in this respect.

While I shall neither probe the implications of MHE for *jus ad bellum*, nor the relationship between *jus ad bellum* and *jus in bello*, I want to draw attention to a third just war framework that is emerging in the philosophical literature on war, *jus post bellum* (Orend 2000). Governing the establishment of just post-war orders, *jus post bellum* includes principles that regulate the prosecution of war crimes and other transgressions of *jus in bello*, as well as general principles of compensation and repair. It is impossible to make sense of these principles without a sound understanding of how responsibility is assigned *in* war, enhanced or not. The arguments of this chapter are, therefore, directly relevant to *jus post bellum*, though I shall not provide a detailed account of post-war justice here.

The third background consideration has to do with the overall permissibility of MHE. When considering the assignment of responsibility in enhanced warfare, one seems to implicitly assume that MHE techniques have been developed. If there were no enhanced soldiers, it would not be possible to deploy them. Without the prospect of deployment, it is not necessary to fret over the assignment of moral responsibility. Thus, the topic of this chapter is directly related to two further questions. Firstly, is it morally permissible to develop MHE techniques? Secondly, is it morally permissible to deploy enhanced soldiers? These two sets of questions are independent from each other. It may be permissible to undertake research that might make MHE possible, but this is not sufficient to show that it is also permissible to deploy enhanced soldiers. There could be many reasons why enhanced combatants should not leave the laboratory.

The chapter's focus on *jus in bello* may be taken to suggest that at least the question of development has been settled: there are enhanced combatants who can (potentially) be deployed, and we are now wondering what this means for the assignment of responsibility in enhanced warfare. However, things are not quite that straightforward. On the one hand, assignments of responsibility in enhanced war seem to have little impact on the moral permissibility of the development of

MHE techniques. There could be many reasons against MHE that have nothing to do with responsibility. For instance, research may be too costly or put subjects under too much stress. On the other hand, the role of responsibility in enhanced warfare is central to the permissibility of research into MHE. For example, if MHE gave rise to pervasive responsibility gaps, thus making the assignment of moral responsibility impossible, it may not be permissible to proceed with the development of MHE techniques.

This takes me to the fourth background consideration. In war, responsibility can be assigned at a number of levels and to a variety of agents, including states. In this chapter, I focus on the individual. Firstly, enhancement techniques will be used on individuals. Secondly, individual agency in war is crucial. The primary disagreement between contemporary just war theorists concerns the extent to which war restricts human agency (Walzer 2006; McMahan 2009). Some are optimistic in this regard, others less so. But even sceptics concede that individual soldiers are responsible for the war crimes they commit (Walzer 2006, p. 39). Hence the chapter needs to consider whether MHE undermines the agency of enhanced soldiers to such an extent that moral responsibility for military acts cannot be assigned to them. In other words, the question is whether enhanced soldiers could cite their enhancement as an exculpating factor in order to be excused for wrongdoing. The next part of the chapter looks at this question in more depth.

Moral Agency and Enhancement in a Post-Nuremberg World

Without agency, assignments of moral responsibility are unjustified. The aforementioned case of a person who becomes a human missile in a freak accident illustrates the point: causal responsibility is not sufficient to be morally responsible for an event. Likewise, being an agent is necessary but not sufficient to be morally responsible. One could imagine a highly sophisticated autonomous robotic weapon that can navigate a complex battlefield without assistance from an operator. But even though such a robot can be described as an artificial agent, it is futile to assign moral responsibility to it. In general, I assume that, in order for moral responsibility to be assignable to a human individual, that individual needs to fulfil three conditions (Cowley 2014, 26–8):

1. *The capacity condition*: In order to be responsible for an event, an agent must have the mental capacity to understand what s/he did, and why s/he is being praised or blamed for an event. In case s/he is blamed for the event, s/he also needs to be aware of the various defences available to him/her (excuse, justification).
2. *The understanding condition*: In order to be responsible for an event, an agent needs to understand a situation that led to the event and his/her role within it. S/he also needs to understand his/her response to the situation.

3. *The control condition*: In order to be responsible for an event, the agent needs control over his/her actions. In particular, s/he must have the capacity not to carry out the act that led to the event.

Of course, what it means to fulfil these three conditions can differ between contexts. Compared to peacetime, war, as already noted, restricts human agency. The historical example of the Nuremberg Trials can assist us with theorising human agency in war. Prior to Nuremberg, soldiers, in order to be exculpated from wrongdoing, had to show that the orders they had received had been duly authorised. Nuremberg introduced two additional criteria, the moral perception and moral choice criteria (May 2005). The former criterion, analogous to the understanding condition, requires that a soldier must have been in possession of the morally relevant facts in order to be responsible for a military act. Recognising that war makes it difficult to acquire the relevant facts, the moral perception criterion merely demands that the soldier must have been capable of distinguishing between legitimate and illegitimate military targets. The moral choice criterion, analogous to the control condition, demands that the soldier must have been able to avoid carrying out a particular military act. If the soldiers can show that s/he lacked knowledge of the relevant moral facts and/or had not been able to have done otherwise, s/he may, depending on the circumstances, be excused for wrongdoing.

It is easy to see how MHE can potentially undermine the moral choice and perception criteria. Enhanced soldiers could argue that enhancement clouded their judgement as to who, or what, was a legitimate or illegitimate target. Or they could argue that enhancement deprived them of their ability *not* to perform a wrongful military act. There is a famous philosophical thought experiment that encapsulates the absence of understanding and control: a person – call him Brian – is kidnapped by a villain and fitted with a brain implant; the villain then remote-controls Brian via the brain implant and orders him to kill other people.[2] Having no moral qualms due to the implant, Brian's ability to kill has been greatly enhanced. He goes about his 'business' efficiently and effectively. Brian poses a lethal threat to others, but fails to fulfil the understanding and control conditions. Consequently, he is causally but not morally responsible for the crimes he commits. Due to the effects of the brain implant, Brian is exculpated from wrongdoing. But is Brian's case a sound way of thinking about the potential perils of MHE? The answer is mixed.

Let me begin by outlining why Brian's case is helpful. It alerts us to a salient issue in the debate on MHE, the question of consent. In Brian's case, no consent is given to the implantation of the device in his brain. Soldiers, by contrast, can have some moral agency in deciding whether to be enhanced or not. To reflect this, it is necessary to distinguish between *consensual* and *non-consensual* MHE. If consent

---

2 Cases such as these have been widely used in discussions of the ethics of self-defence. Brian is an innocent attacker, who is not morally responsible for the threat he poses. On this issue, see McMahan (1994) and Otsuka (1994).

is given by a soldier, there does not seem to be a problem with assigning moral responsibility to him/her for his/her actions during enhanced war. Note that this is even true if the soldier's agency is compromised as the result of the enhancement. If, at t_2, the enhanced soldier's moral perception of a situation, or his/her ability not to perform an act, is undermined by the enhancement, his/her consent at t_1 means that one can still assign (retrospective) responsibility to him/her at t_3. The criminal law serves as a useful guide here. Voluntary intoxication is no excuse for causing grievous bodily, or even lethal, harm to others.[3] A driver who gets drunk in the pub and runs over a pedestrian on his way back is usually not exculpated from wrongdoing because of his drunkenness.

Critics can point out that, while we can assume that the driver got drunk out of his own free will, the military, being a hierarchical institution, has the means to put its members under pressure to undergo enhancement. Suppose that Ben is offered the following choice. If Ben accepts enhancement in order to be able to participate in a difficult mission, he will be promoted. If Ben does not undergo enhancement and does not participate in the mission, his career will not advance any further. I do not think that Ben, if he consents to enhancement, would be exculpated from wrongdoing because his superior put pressure on him. Even an extreme case of duress does usually not excuse wrongful killing (Rodin 2002, p. 171). The situation experienced by Ben falls short of duress. His career is on the line, not his life. Mild pressure falling short of duress is not sufficient to show that assignments of moral responsibility to Ben are not justified.

However, manipulation can take more subtle forms than just described. Ben might consent to be enhanced because Ben's superiors left him in the dark about the impact of MHE on his agency. The behaviour of Ben's superiors, in my view, constitutes a violation of duty of care towards Ben, as well as those who might potentially be harmed illegitimately by Ben in his enhanced state. I return to the concept of due care later. For now, I want to stress that the behaviour of Ben's superiors, though morally wrong, does not lead to a responsibility gap. True, it might not be possible to assign moral responsibility to Ben for non-compliance with *jus in bello* norms. But just as the villain who implanted the chip in Brian's brain is responsible for the killings Brian commits, moral responsibility can be assigned to Ben's superiors. Ben's example shows that, in the context of the MHE debate, it is necessary to operate with a more demanding notion of consent, *informed* consent. Soldiers must be aware of the potential effects of MHE on their agency, otherwise their consent only has limited normative weight. Still, with or without informed consent, there is no responsibility gap.

Highlighting the issue of informed consent, the above illustrates the helpfulness of Brian's case in the present debate on MHE. Yet there also is a sense in which this thought experiment obscures the issue. Brian's loss of agency is complete and, most importantly, intended by another party, the villain. In reality, the implications

---

3 I am grateful to Laurence Lustgarten for discussing the status of intoxication in criminal law with me.

of MHI for moral agency are more complex. We can differentiate between three different scenarios here.

Firstly, MHE may not always have a detrimental effect on moral agency. Imagine Chris has been enhanced so he can run faster. Suppose that this enhancement does not have any impact on Chris' ability to assess a particular situation and control his actions. Chris can simply run faster now, that is all. At least insofar as the assignment of moral responsibility is concerned, the fact that Chris' ability to run faster has come about through biomedical enhancement, rather than physical training alone, does not make a moral difference. Chris is not more or less responsible for compliance, or non-compliance, with *jus in bello* norms than if he had trained very hard.

Secondly, MHE may sometimes enhance, rather than diminish, moral agency. Suppose Carl is given an anti-anxiety drug before he is sent on a dangerous mission. Fear can cloud an individual's judgement by diminishing his/her moral perception. Innocent individuals, who are illegitimate targets, may suddenly appear as legitimate targets. The infamous My Lai massacre serves as a potent reminder of this. Surely, we would judge Carl differently to a young recruit who has not been given the anti-anxiety drug. The point, however, is that the administration of the anti-anxiety drug has few repercussions for the *assignment* of moral responsibility to Carl. Rather, it means that we can apply a higher *standard* of responsibility to Carl than would be justified in the case of the young recruit. But this does not change the fact that we assign responsibility to both. It is, therefore, important to distinguish between assignments of responsibility and standards of responsibility. MHE often does not impact on the former, but has repercussions for the latter. The problem is not that MHE leads to responsibility gaps, but that it necessitates a rethinking of standards of responsibility in the armed services.

Thirdly, we need to distinguish between intentional and non-intentional impairments of agency. Imagine that Jack and John, operators of a missile defence system, have been given a pill that increases their alertness. As a result, Jack and John can now concentrate on what would otherwise be a boring task for prolonged periods of time – staring at the system's radar screen to identify potentially hostile aircraft. Unfortunately, the pill makes Jack and John more open to persuasion.[4] As soon as Jack spots an object on the radar screen, he exclaims that it must be hostile, an assessment John immediately finds persuasive. John looks at his screen and seconds Jack's assessment. Jack, feeling vindicated, reports a hostile object to his and John's superiors. Jack and John have become better at their task – concentrating on the radar screen for a long time – but their moral perception – is the object behaving in hostile ways? – has been impaired.

Jack's and John's case differs from Brian's. This is because the impairment of Jack's and John's moral agency is a side effect of MHE. The intended effect is that Jack and John are able to concentrate for longer, not that they persuade each other more easily. Reasoning counterfactually, Jack's and John's superiors

---

4   I am grateful to Anders Sandberg for discussing this case with me.

would still administer the pill, even if it would *not* make Jack and John more susceptible to persuasion.[5] Jack and John's superiors are interested in fast *and* accurate assessments of potentially hostile objects. At the moment Jack and John provide fast assessments of objects, but these are not always accurate. But accuracy is certainly desirable in a military context in order to ensure compliance with the principles of *jus in bello*. The impairment of Jack and John's agency is a foreseen side effect of the administration of the pill, but not its intended effect. In Brian's case, by contrast, the villain intends to deprive Brian of his moral agency. Reasoning counterfactually, the villain would not implant the control device in Brian's brain if it did not turn Brian into his willing tool, or a 'killing machine'.

Non-intentional impairments of moral agency, I contend, may be morally permissible. The loss of agency is a side effect, and the importance of moral agency is not denied by the party that practises MHE. If it was possible to preserve agency by supressing the unintended side effect of MHE, there is no reason not to do so. However, the use of MHE in order to intentionally deprive soldiers of their moral agency is impermissible, regardless of whether the soldier consents to it or not. There are two reasons for this. Firstly, doing so would be a regressive step into a pre-Nuremberg world. Nuremberg recognised that soldiers are not automatons, but human persons who are capable of exercising moral agency, even in war. Secondly, there are quasi-Kantian reasons why the intentional removal of agency is morally impermissible. A quasi-Kantian position would demand that we must not treat other persons as a means to an end but also always as ends-in-themselves. To be sure, the hierarchical structures of authority and obedience of the military might trouble Kantians. But let us assume that some restrictions of moral agency – and, in particular, autonomy in Kant's case – can be morally justified, especially in terms of a security crisis. What cannot definitely not be justified – crisis or no crisis – is turning individuals into willing tools without *any* moral agency. Note that, on the quasi-Kantian view, the soldier's 'consent' to such treatment is not morally valid: human agents are not permitted to *completely* surrender their moral agency, effectively treating themselves as means at the ready disposal of others.

The question is whether the distinction between intentional and non-intentional impairments of moral agency has implications for the assignment of responsibility in enhanced warfare. The answer is ambiguous. On the one hand, it does not. If

---

5   I follow the Doctrine of Double Effect here, which forms an important component of non-consequentialist ethics. This doctrine distinguishes between what an agent intends and what he foresees. Assuming that many acts have morally positive as well as negative consequences, the doctrine contends that an act is only permissible if its positive consequences are intended by the agent who performs the act, while the negative consequences are merely foreseen by that agent. The permission to cause an unintended side effect is usually further restricted by a proportionality criterion. The harm caused by the foreseen side effect must not be excessive to the good achieved via the intended effect. The literature on the Doctrine of Double Effect is vast, see Cavanaugh (2006).

soldiers make an informed choice to surrender some, or all, of their moral agency by consenting to MHE, responsibility can still be assigned to them. As indicated above, if they do not consent to the loss of agency, responsibility can be assigned to their superiors. On the other hand, the intentional impairment of agency via MHE broadens the scope of assignments of responsibility. So far, the discussion has focused on the assignment of moral responsibility to individual soldiers and their superiors. In case of intentional impairment, moral responsibility also should be assigned to those who develop relevant MHE techniques. While I think that there is a *prima facie* permission to participate in weapons research, participation in the development of MHE techniques that consciously and intentionally undermine moral and legal standards is impermissible. Those who deliberately develop such techniques will have to share some of the moral blame if enhanced soldiers fail to comply with *jus in bello*.

In sum, the above shows that worries about potential responsibility gaps are, in the case of MHE, unwarranted. Responsibility for wrongdoing in the course of enhanced warfare can either be assigned to enhanced soldiers or their superiors. In many cases, responsibility can be assigned to both. However, even though MHE does not pose fundamentally new challenges for the assignment of responsibility in enhanced warfare, the above analysis provides important insights. Firstly, it highlights the significance of informed consent in enhanced warfare. Secondly, it shows that certain forms of MHE are morally impermissible. At the core of assignments of moral responsibility lie wider issues about the normative significance of moral agency. MHE techniques that deny moral agency to soldiers are impermissible. That said, as we saw in the Jack and John example, non-intentional impairments of agency, though potentially morally permissible, are not unproblematic either. Non-intentional impairments raise important questions with regard to assignments of prospective responsibility. I offer a brief discussion of this topic in the next part of the chapter.

Prospective Responsibility and Standards of Due Care

The above arguments are most relevant for a retrospective understanding of moral responsibility. A soldier who has undergone enhancement did not comply with *jus in bello* and we want to know whether it is possible to assign moral responsibility to him/her for his/her behaviour. The reference to the Nuremberg Trials reinforces the retrospective dimension: criminal trials are typically concerned with past wrongdoing. Punishment, the focus of criminal law, is reactive, not proactive. Yet, as indicated in the introduction to the chapter, responsibility also can be forward-looking. Assignments of prospective responsibility usually involve the creation of specific roles, and corresponding obligations, for agents. The example of Jack and John, the enhanced operators of a missile defence system, helps to illustrate the point. As a result of the concentration-enhancing pill, you recall, Jack and John have become more susceptible to persuasion. They are, in their enhanced state, much better at tracking potentially hostile objects than unenhanced operators. However, while their assessments are fast, there is no guarantee that they are accurate. Because of this, it would be catastrophic if Jack and John's superiors believed every threat assessment the two operators made.

One way to make enhancement safe in this context consists in creating appropriate roles not only for Jack and John, but also their superiors. We do not want to wait until it is too late, and then assign retrospective responsibility to Jack and John for wrongdoing. Rather, we want to prevent wrongdoing in the first place. Jack and John's superiors must thus be aware that the two operators provide them with fast but not always accurate assessments. In order to prevent the application of force to illegitimate targets, Jack and John's superiors are obliged to provide a separate threat assessment. The fact Jack and John spot new objects early on might be desirable because it affords more time to other agents to undertake extensive threat assessments. The safe operation of the missile defence system would involve the creation of role specific duties, and corresponding command and control structures, for those under whose command Jack and John serve.

This example, and the idea of prospective responsibility in general, gives rise towards a wider point about standards of due care. The use of military force, whether through enhanced soldiers or not, always creates certain risks. These risks need to be mitigated, and this can only be done by developing a sound standard of due care. Such a standard identifies the particular risks arising in a specific domain. It then finds ways to ensure that these risks remain reasonable. It needs to be stressed that the use of military force can never be 100 per cent safe. Similarly, a perfect safety threshold cannot be the aim of standards of due care. Nevertheless, sound standards of due care can significantly reduce risks. In Jack and John's case, direct supervision by a superior officer and separate assessment procedures for the threat potential of incoming objects should form part of an acceptable standard of care.

It is probably trivial to say that MHE can have great advantages. But these must be balanced against potential risks. Here the argument comes full circle. Firstly, the development of a sound standard of care that assigns role-specific responsibility prospectively may be crucial for the moral permissibility of deploying enhanced soldiers, if not the development of MHE techniques. Secondly, an appropriate standard of due care will also have repercussions for the assignment of retrospective responsibility. Retrospective responsibility is not only assignable to individual soldiers who are causally involved in wrongdoing. It is also assignable to those who neglected their role-specific duties of care, even if they were not directly causally involved in an impermissible military act. Thus, although we can conceptually distinguish between retrospective and prospective responsibility, the two concepts are closely related.

Conclusion

In this chapter, I have argued that the worry that MHE gives rise to responsibility gaps is unfounded. Despite this, an engagement with MHE raises a number of interesting questions for responsibility. Firstly, just war theory tends to foreground retrospective responsibility, rather than prospective assignments of responsibility. This is unfortunate. As this discussion has shown, prospective responsibility is also an important issue to consider in the MHE debate and beyond. Secondly, and directly related to the preceding point, the permissibility of MHE depends partly on the formulation of a sound standard of due care, in which role-specific responsibilities are assigned to those who work with or have authority over enhanced soldiers. Thirdly, the assignment of responsibility directly leads to wider questions about *informed* consent to MHE. I argue that the informed consent of soldiers to enhancement procedures is crucial. Even if MHE impairs the moral agency of a soldier, responsibility for wrongdoing can still be assigned to the soldier if s/he gave his/her informed consent to the treatment. Finally, the development of enhancement procedures that intentionally undermine the moral agency of soldiers must be prohibited. Responsibility for wrongdoing caused by deliberate impairments of agency is not just assignable to individuals who command enhanced soldiers, but also researchers who developed relevant enhancement techniques. These four points show that, while a responsibility gap misses the mark on the MHE debate, there are nonetheless a number of important issues with regard to responsibility that warrant further research as the debate on MHE moves forward.

References

Agar, N. 2014, *Truly Human Enhancement: A Philosophical Defense of Limits*. Cambridge, MA: MIT Press.

Blank, R.H. 2013, *Intervention in the Brain: Politics, Policy and Ethics*, Cambridge, MA: MIT Press.

Buchanan, A. 2011, *Beyond Humanity*. Oxford: Oxford University Press.

Cavanaugh, T.A. 2006, *Double Effect Reasoning: Doing Good and Avoiding Evil*. Oxford: Oxford University Press.

Cowley, C. 2014, *Moral Responsibility*. Durham, UK: Acumen.

Friscolanti, M. 2005, *Friendly Fire: The Untold Story of the U.S. Bombing that Killed Four Canadian Soldiers in Afghanistan*. Mississauga, ON: Wiley Canada.

Grossman, D. 2009, *On Killing: The Psychological Cost of Learning to Kill in War and Society*. New York: Back Bay Books.

Harris, J. 2007, *Enhancing Evolution: The Ethical Case for Making Better People*. Princeton, NJ: Princeton University Press.

McMahan, J. 2009, *Killing in War*. Oxford: Oxford University Press.

McMahan, J. 1994, 'Self-Defense and the problem of the innocent attacker', *Ethics*, 104(2): 193–221.

May, L. 2005, *Crimes Against Humanity: A Normative Approach*. Cambridge: Cambridge University Press.

Moreno, J.D. 2012, *Mind Wars: Brain Science and the Military in the 21st Century*. New York: Bellevue Literary Press.

Orend, B. 2000, *War and International Justice: A Kantian Perspective*. Waterloo, ON: Wilfried Laurier University Press.

Otsuka, M. 1994, 'Killing the innocent in self-defense', *Philosophy and Public Affairs*, 23(1): 74–94.

Persson, I. and Savulescu, J. (2012) *Unfit for the Future: The Need for Moral Enhancement*. Oxford: Oxford University Press.

Rodin, D. 2002, *War and Self-Defense*. Oxford: Oxford University Press.

Sparrow, R. 2007, 'Killer Robots', *Journal of Applied Philosophy*, 24(1): 62–77.

Taylor, K. 2012, *The Brain Supremacy: Notes from the Frontiers of Neuroscience*. Oxford: Oxford University Press.

Walzer, M. 2006, *Just and Unjust Wars: A Moral Argument With Historical Illustrations*. New York: Basic Books.

Chapter 12

Collective Responsibility for the Robopocalypse

Seumas Miller

Science fiction movies, such as the *Terminator* series, have accustomed us to images of armed, computerised robots led by leader robots fighting wars against human combatants and their human leaders. Moreover, by virtue of developments in artificial intelligence, the robots have superior calculative and memory capacity; after all, they are computers. In addition, robots are utterly fearless in battle; they do not have emotions and care nothing for life over death. Does the human race, then, face robopocalypse? The short answer is in the negative. Computers, robotic or otherwise, are not minded agents, steadfast intentional stances toward them notwithstanding (Dennett 1987). Rather these images are fanciful anthropomorphisms of machines; and the military reality is quite different. Nevertheless, the spectre of robopocalypse persists, especially in the context of new and emerging (so-called) autonomous robotic weaponry. Consider, for example, the Samsung stationary robot which functions as a sentry in the demilitarised zone between North and South Korea. Once programmed and activated, it has the capability to track, identify and fire its machine guns at human targets without the further intervention of a human operator. Predator drones are used in Afghanistan and the tribal areas of Pakistan to kill suspected terrorists. While the ones currently in use are not autonomous weapons, they could be given this capability, in which case, once programmed and activated, they could track, identify and destroy human and other targets without the further intervention of a human operator. Moreover, more advanced autonomous weapons systems, including robotic ones, are in the pipeline.

In this chapter I explore the implications of autonomous robotic weapons, and related military weaponry, for the individual and collective moral responsibility of human beings engaged in war. Do such weapons necessarily compromise the moral responsibility of human combatants and their leaders and, if so, in what manner and to what extent? In order to answer these questions we first need serviceable theoretical descriptions of the key notions of war and military necessity (the first section), and individual and collective moral responsibility (the second section). In the third and final section, I turn directly to the questions arising for individual and collective moral responsibility in respect of autonomous robotic weaponry. Importantly, I provide what I refer to as the moral ramification argument. The conclusion of this argument is that it is highly improbable that moral *jus in bello* principles of military necessity, discrimination and proportionality could ever be programmed in to robots.

War, Collection Action and the Principle of Military Necessity

Wars are ongoing, serious, armed conflicts between the armed forces of political entities, such as nation-states, revolutionary groups and terrorist groups. Such armed conflicts include civil wars, wars of liberation and non-conventional wars between state actors and terrorist groups. An armed force in the relevant sense is a collective entity: (i) comprised of combatants with task-defined roles; (ii) with a command and control structure; (iii) which reproduces itself, for example by way of recruitment and training; (iv) engaged in a collective enterprise, namely, armed conflict against another armed force; (v) to realise a collective military end (or ends), for example incapacitate the enemy armed force; (vi) which collective military end is ultimately in the service of a collective political end(s) of the political entity of which the armed force is the military wing.

Waging war is typically morally justified by recourse to some notion of collective self-defence, for example defence of the nation-state against the armed aggression of another nation-state or of a non-state actor such as a terrorist group (Walzer 1977). This ultimate end of collective self-defence and, relatedly, winning the war is necessarily underspecified prior to its realisation. For example, the US did not know when it declared war on Japan as a result of the Japanese attack on Pearl Harbour that victory over Japan would ultimately result from dropping atom bombs on Nagasaki and Hiroshima. Moreover, the ins and outs of the evolving route leading to victory is also necessarily underspecified prior to its actually being taken; after all, it largely turns on what the enemy does, including by way of response to one's own armed attacks. So war is quite unlike programming in the final destination to a robot-driven car with a detailed and fixed roadmap or, for that matter, the flight path to a computer-controlled jet aircraft. Nor is it even like playing a game such as chess, albeit it is analogous in some ways. For unlike in war, in chess there is a single, definite, unchanging and mutually known 'theatre of war' (the chessboard), a resource base which cannot reproduce itself (the chess pieces), a sharply defined set of rules and contexts of application and a fixed, finite and knowable (at least in principle) set of possible moves and counter-moves.

The actual conduct of war is governed by moral principles (the so-called *jus in bello* of just war theory), notably the principles of (1) military necessity, (2) proportionality and (3) discrimination.[1] As will become evident, these are quite unlike the sharply defined rules and contexts of application in chess. For the moment I note that these principles have to be applied in very different military contexts, for

---

1   There are various different possible formulations of and complications arising from these moral principles. For example, I will be concerned with proportionality as it pertains to civilian deaths. See Miller (2009).

example conventional theatres of war and counterterrorism operations, and that, as I argue below, their application is *radically context dependent* – so the conditions in which they ought to be applied cannot be comprehensively specified in advance of those conditions coming into existence. Importantly, unlike in the case of law enforcement, these principles apply at the collective level, as opposed to merely at the individual level. So the context of any or, at least, most applications of these principles is *multi-levelled*. What do I mean by the collective level(s)?

Consider the following scenario involving a military organisation engaged in battle. It involves what I refer to as a *multilayered structure of joint actions* (Miller 2010, p. 48). If two or more individuals perform a joint action, then each of them intentionally performs an individual action (or omission), but does so with the (true) belief that in so doing they will jointly realise an end which each of them has and, in these circumstances, would not have if the other did not since neither could readily realise the end on his or her own (Miller 1992, 1995). For example, two gunners mounting a large gun on its turret are engaged in joint action. Suppose at an organisational level a number of discrete joint actions ('actions') are severally necessary and jointly sufficient to achieve some collective end. Thus the 'action' of the mortar squad destroying enemy gun emplacements, the 'action' of the flight of military planes providing air cover and the 'action' of the infantry platoon taking and holding the ground might be severally necessary and jointly sufficient to achieve the collective end of winning the battle; as such, these 'actions' taken together constitute a joint action. Call each of these 'actions' level two 'actions', and the joint action that they constitute also a level two joint action. From the perspective of the collective end of winning the battle, each of these level two 'actions' is an individual action that is a component of a (level two) joint action: the joint action directed to the collective end of winning the battle.

However, each of these level two 'actions' is already in itself a joint action with component individual actions; and these component individual actions are severally necessary (let us assume this for purposes of simplification, albeit it is unlikely that every single action would in fact be necessary) and jointly sufficient for the performance of some collective end. Thus the individual members of the mortar squad jointly operate the mortar in order to realise the collective end of destroying enemy gun emplacements. Each pilot, jointly with the other pilots, strafes enemy soldiers in order to realise the collective end of providing air cover for their advancing foot soldiers. Further, the set of foot soldiers jointly advance in order to take and hold the ground vacated by the members of the retreating enemy force.

Accordingly, at level one there are individual actions directed to three distinct collective ends: the collective ends of (respectively) destroying gun emplacements, providing air cover and taking and holding ground. So at level one there are three joint actions, namely, the members of the mortar squad destroying gun emplacements, the members of the flight of planes providing air cover and the members of the infantry taking and holding ground. However, taken together these three joint actions constitute a single level two joint action *at the collective level* (so to speak). The collective end of this level two joint action is to win the battle; and from the perspective of this level two joint action, and its collective end, these constitutive actions are (level two) individual actions.

Nor is this the end of the matter for, as we all know, any given battle is merely a phase element in the overall war. So there are further collective levels governed by, for example, the collective end of winning the war, as opposed to merely winning one of the battles. Perhaps winning the war is describable as a level three joint action.

The point to be stressed now is that the principle of military necessity, in particular, but also the principles of proportionality and discrimination, apply at the various conceptually distinct collective levels (for example the level of a battle or ongoing war fought by a military organisation), and not simply at the level of an individual combatant's lethal action considered as a discrete, self-contained action (for example the necessity to kill an enemy combatant who will otherwise kill oneself). Accordingly, the context for the application of these moral principles is a multi-level (individual and collective end) context. Let me explain.

In essence, the principle of military necessity ultimately pertains to the long-term, necessarily underspecified collective end of winning the war which generates in turn a nested, dynamic, series of medium and short-term collective ends, such as winning particular battles or firefights. These short- and medium-term collective ends are means to the long-term collective end of winning the war, albeit means in need of further specification, adjustment or even abandonment in the light of the responses to them of the enemy armed forces. Accordingly, the principle of military necessity is to be understood, firstly, in short-/medium-/long-term means/end, that is, diachronic, terms. Something is necessary in this sense if, comparatively speaking, it is both an efficient and effective means to an end and there is no obviously superior means available. If it is the *only* means then it is *strictly* necessary. However, this is frequently not the case and so to this extent 'necessity' is correspondingly less strict. Secondly, the strength of the necessity to deploy a given quantum of lethal military force in (say) the context of a battle turns in large part on the moral weight to be accorded to the winning of that battle in light of its likely contribution to the ultimate (necessarily underspecified) collective end of winning the war (and, of course, the somewhat indeterminate moral weight to be attached to the latter). In the case of a crucial battle in the context of a war of collective self-defence, the military necessity to deploy a large quantum of lethal military force might be both strong (there is much at stake) and strict (it is the only available means). What of the principles of proportionality and discrimination?

These principles are obviously also to be applied at all collective and individual levels: whether it is a brief one-combatant-to-one-enemy-combatant exchange of fire, a firefight involving multiple combatants on both sides, a battle or the war as a whole that is under consideration, it is morally impermissible to intentionally kill innocent civilians, or put their lives at unnecessary risk or knowingly cause disproportionate large numbers of civilian deaths. Naturally, what is at stake at each of these different levels, including the quantum of lives, can vary greatly but this does not affect the applicability of the principles.

The principles of military necessity, discrimination and proportionality are logically interdependent; one cannot be correctly applied without attending to the requirements of the others. Let me explain.

Roughly speaking, the principle of discrimination forbids intentional targeting of innocent civilians[2] and, also, foreseeably and avoidably putting their lives at unnecessary risk. The latter clause conceptually implicates the principle of military necessity; a risk to civilians is unnecessary if the use of lethal military force which constitutes this risk is not militarily necessary. So the principles of military necessity and discrimination are logically interdependent. Moreover, as we saw above, both principles must be applied at all individual and collective levels. Since these levels are interconnected by virtue of nested collective ends, the application of the principle of discrimination may well be a complex matter necessarily involving taking into account: (i) the risks to civilians at these various levels and (possibly) adjudicating between them, and; (ii) military necessity at these various levels and (possibly) adjudicating between them, and; (iii) adjudicating between (i) and (ii). For example, pursuing tactic A (aerial bombing) to realise the collective end of winning a battle might lead to many more civilian casualties in this present battle than pursuing tactic B (taking and holding ground without aerial bombing). However, pursuing A might be a more efficient and effective means of decisively winning the battle (because, say, of the much heavier enemy casualties inflicted prior to the enemy's retreat) and might, therefore, reduce the number of future civilian casualties in future battles joined in further pursuit of the collective end of winning war. What of the principle of proportionality?

The principle of proportionately arises in contexts in which both the principle of military necessity and the principle of discrimination are applicable.[3] Roughly speaking, it requires that that the quantum of (unintended) civilian deaths resulting from the deployment of lethal military force should not be disproportionate to the strategic value, and corresponding moral weight, of the collective military ends to be realised by that deployment. As such, the principle of proportionality is logically interdependent with both the principle of military necessity and the principle of discrimination. Moreover, as we saw above, the principle of proportionality applies at both the individual and collective levels. So the application of the principle of proportionality is complex in the manner of the other two principles.

---

2   Arguably, the component clause of the principle of discrimination, namely, the impermissibility of intentionally killing innocent civilians is logically independent of its second clause and of the other principles. This does not affect my argument. The principle of discrimination also applies to kind of weaponry uses. For example, biological weapons are indiscriminate.

3   For a recent useful discussion of the principle of proportionality in relation to individual self-defence see Uniacke (2011).

This combination of logical interdependence between the three *jus in bello* principles and their applicability at all interconnected individual and collective levels in the overall context of a just war waged in collective self-defence gives rise to the phenomenon I refer to as *moral ramification* and the associated need for complex decision-making of the kind described above. In short, one cannot simply apply one of these principles in a discrete, self-contained context (for example, proportionality given the likelihood of heavy civilian casualties in a firefight), without taking into account the other principles and other contexts at other levels (for example, the military necessity to win the battle in which the firefight is an important constitutive element).

Finally, I note that the moral considerations that arise from collective military ends at the collective level often outweigh, or otherwise render irrelevant, the moral considerations that arise at the individual level. In this respect the deployment of lethal force by the military in war is quite different from the use of lethal force by police in law enforcement (Miller and Blackler 2005, pp. 61–82). For example, in war combatants are morally (and legally) permitted to ambush and kill enemy combatants in the service of (say) a medium-term collective military end of winning a battle, notwithstanding that it was not necessary to do so in order to preserve their own lives (or the lives of any civilians). By contrast, it is not morally permissible for police officers to ambush and kill armed criminal offenders. Again, in war it might be morally permissible to put the lives of a dozen innocent civilians at risk in order to kill a single combatant if, by the lights of (say) a long-term collective military end, the combatant in question was of sufficiently high value, for example a high-ranking officer central to the enemy's war effort. By contrast, it would not be morally permissible for a police officer to put the lives of a dozen innocent civilians at risk by shooting at an important organised crime figure in order to prevent his escape.

Individual and Collective Moral Responsibility

Collective moral responsibility is a species of moral responsibility. Here we need to distinguish *moral* responsibility (including collective moral responsibility) from *causal* responsibility.[4] A person or persons can inadvertently cause a bad outcome without necessarily being morally responsible for so doing. Moral responsibility typically requires not only causal responsibility but also an intention to cause harm or the knowledge that one's action will or may well cause harm (or, at the very least, that one ought to have known that it might). We also need to distinguish moral responsibility for actions and moral responsibility for omissions, and retrospective from prospective moral responsibility. All these distinctions in respect of individual moral responsibility are mirrored in the case of collective

---

    4   I make many of these points in previous publications. See, for example, Miller (2006) and Miller (2014).

moral responsibility. Collective moral responsibility is the moral responsibility that attaches to structured groups, such as armed forces, as well as unstructured groups for their morally significant actions and omissions, including in the application of the principles of military necessity, discrimination and proportionality.

Elsewhere I have elaborated and defended a relational account of collective moral responsibility; specifically, that of collective responsibility as joint responsibility (Miller 2006). On this view, collective responsibility is responsibility arising from joint actions and omissions. On this view of collective responsibility as joint responsibility, collective responsibility is ascribed to individual human beings only, albeit jointly. Each member of the group is individually morally responsible for their contributory action and also for the outcome of the set of actions. However, each is individually responsible for that outcome, *jointly with the others*; hence the conception is relational in character.

I have argued that collective moral responsibility is to be understood as joint moral responsibility: the joint moral responsibility of individual human actors engaged in morally significant joint actions. I have further argued that the notion of joint action can be enriched so as to encompass organisational action; it does so by way of multilayered structures of joint action. The upshot of this analysis is that individual human actors are, at least in principle, collectively (jointly) morally responsible for morally significant organisational action and, in particular, joint military activity.[5] Accordingly, given that 'the action' of (say) an armed force in winning a battle is to be understood as a multilayered structure of joint actions, and given this joint action is morally significant, then, other things being equal, the various institutional role occupants, for example the combatants and their commanders, who participate in it are collectively (jointly) morally responsible for the successful joint action and for its foreseeable untoward outcomes, for example civilian casualties.

Here it is important to note that within the set of individuals who are collectively morally responsible for some outcome, the degree of individual responsibility that some have (jointly with others) might be greater than the degree of individual responsibility that those others have, for example commanders will typically have a higher degree of individual responsibility than their subordinates. Indeed, if the contribution of some individuals is minute and they are only very indirectly connected to some morally significant outcome, then their degree of moral responsibility may well diminish to the point of non-existence.

Moreover, these various levels and nested structures of morally significant institutionally based joint activity give rise to an associated structure of causally and, given the structure of nested individual and collective ends, morally interconnected *centres* of individual and collective responsibility. One such centre might be the designers and manufacturers without whom there would be

---

5 This theoretical standpoint is not to be confused with the view that organisations and other collective entities can be reduced to the individual human organisational actors and their individual actions. The latter view is surely incorrect.

no weapons. They are directly collectively morally responsible for providing the means for others to wage war with all its foreseeable death and destruction, but also with the possibility of a morally justified victory. Another is the senior military and political leadership who jointly decided to go to war and who, therefore, may well have an important share of the collective moral responsibility for the foreseeable (morally good, let us assume) consequences of the war in question. Still another are the commanders who, in the overall context of the pursuit of the collective end of winning the war in questions, put in place given rules of engagement (ROE) in a specific theatre of war, and the combatants who comply with those ROE. These human agents are collectively morally responsible, other things being equal, for the consequences of those ROE in that theatre, albeit subordinates may have diminished individual responsibility relative to commanders. Then there are the relevant officers, combatants, intelligence personnel and weapons programmers/ activators/operators of a given weapons system involved in, or relied upon, in a given firefight. These are collectively morally responsible, other things being equal, for the outcomes (good and bad) of the use of this weaponry in this firefight.

Further, in some cases of collective moral responsibility no one is *fully* morally responsible for the adverse outcome; rather each has a share, so to speak, of the collective moral responsibility in question. On the conception of collective moral responsibility as joint moral responsibility, each member of the group in the centre in question must have *some* degree of moral responsibility (jointly with the others).[6] Naturally, multiple individuals could be collectively *causally* responsible for some adverse outcome without any individual having any *moral* responsibility (notwithstanding his or her individual causal responsibility).[7]

Autonomous Robotic Weaponry and Human Moral Responsibility

Autonomous weapons are weapons system which, once programed and activated by a human operator, can – and, if used, do in fact – identify, track and deliver lethal force without further intervention by a human operator. By 'programmed' I mean, at least, that the individual target or type of target has been selected and programmed into the weapons system. By 'activated' I mean, at least, that the process culminating in the already programmed weapon delivering lethal force

---

6   For arguments against collectivist theories of collective moral responsibility which allow the possibility of collective moral responsibility without any individual moral responsibility, or with collective moral responsibility above and beyond aggregate (and/or joint) moral responsibility, see Miller (2007), Miller and Makela (2005).

7   A final point is that the members of an armed force engaged in armed conflict with an enemy armed force may have shared, but not joint (or, therefore, not collective in my sense) moral responsibility for some untoward outcome caused by their fighting, for example the exodus and subsequent impoverishment of the civilian occupants of some location turned into a theatre of war by their fighting.

has been initiated. This weaponry includes weapons used in non-targeted killing, such as autonomous anti-aircraft weapons systems used against multiple attacking aircraft or, more futuristically, against swarm technology (for example multiple lethal miniature attack drones operating as a swarm so as to inhibit effective defensive measures); and ones used or, at least, capable of being used in targeted killing (for example a predator drone with face-recognition technology and no human operator to confirm a match).

We need to distinguish between so-called 'human *in*-the-loop', 'human *on*-the-loop' and 'human *out-of*-the-loop' weaponry. It is only human out-of-the-loop weapons that are autonomous in the required sense. In the case of human-*in*-the-loop weapons, the final delivery of lethal force (for example by a predator drone) cannot be done without the decision to do so by the human operator. In the case of human *on*-the-loop weapons, the final delivery of lethal force can be done without the decision to do so by the human operator; however, the human operator can override the weapon system's triggering mechanism. In the case of human *out*-of-the-loop weapons, the human operator cannot override the weapon system's triggering mechanism; so once the weapon system is programmed and activated there is, and cannot be, any further human intervention.

The lethal use of a human-*in*-the-loop weapon is a standard case of killing by a human combatant and, as such, is presumably, at least in principle, morally permissible. Moreover, other things being equal, the combatant is morally responsible for the killing. The lethal use of a human-*on*-the-loop weapon is also in principle morally permissible. Moreover, the human operator is, perhaps jointly with others, morally responsible, at least in principle, for the use of lethal force and its foreseeable consequences. However, these two propositions concerning human on-the-loop weaponry rely on the following assumptions:

1. The weapon system is programmed and activated by its human operator *and either*;
2. (a) On each and every occasion of use the final delivery of lethal force can be overridden by the human operator and; (b) this operator has sufficient time and sufficient information to make a morally informed, reasonably reliable judgement whether or not to deliver lethal force *or*;
3. (a) On each and every occasion of use the final delivery of lethal force can be overridden by the human operator and; (b) there is no moral requirement for a morally informed, reasonably reliable judgement on each and every occasion of the final delivery of force.

A scenario illustrating (3)(b) might be an anti-aircraft weapons system being used on a naval vessel under attack from a squadron of manned aircraft in a theatre of war at sea in which there are no civilians present.[8]

---

8   There are various other possible such scenarios. Consider a scenario in which there is a single attacker on a single occasion in which there is insufficient time for a reasonably

What of human *out-of*-the-loop weapons?[9] Consider the following scenario which, I contend, is analogous to the use of human out-of-the-loop weaponry. There is a villain who has trained his dogs to kill on his command and an innocent victim on the run from the villain. The villain gives the scent of the victim to the killer-dogs by way of an item of the victim's clothing and then commands the dogs to kill. The killer-dogs pursue the victim deep into the forest and now the villain is unable to intervene. The killer-dogs kill the victim. The villain is legally and morally responsible for murder. However, the killer-dogs are not, albeit they may need to be destroyed on the grounds of the risk they pose to human life. So the villain is morally responsible for murdering the victim, notwithstanding the indirect nature of the causal chain from the villain to the dead victim; the chain is indirect since it crucially depends on the killer-dogs doing the actual physical killing. Moreover, the villain would also have been legally and morally responsible for the killing if the 'scent' was generic and, therefore, carried by a whole class of potential victims, and if the dogs had killed one of these. In this second version of the scenario, the villain does not intend to kill a uniquely identifiable individual,[10] but rather one (or perhaps multiple) member(s) of a class of individuals.[11]

By analogy, human out-of-the-loop weapons – so-called 'killer-robots' – are not morally responsible for any killings they cause (Sparrow 2007).[12] Consider the case of a human in-the-loop or human-on-the-loop weapon. Assume that the programmer/activator of the weapon and the operator of the weapon at the point of delivery are two different human agents. If so, then other things being equal they are jointly (that is, collectively) morally responsible for the killing done by the weapon (whether it be of a uniquely identified individual or an individual qua member of a class).[13] No-one thinks the weapon is morally or other than causally responsible for the killing. Now assume this weapon is converted to a human *out-of*-the-loop weapon by the

---

reliable, morally informed judgement. Such scenarios might include ones involving a kamikaze pilot or suicide bomber. If autonomous weapons were to be morally permissible, the following conditions at least would need to be met: (i) prior clear-cut criteria for identification/delivery of lethal force to be designed-into the weapon and used only in narrowly circumscribed circumstances; (ii) prior morally informed judgement regarding criteria and circumstances, and; (iii) ability of operator to override system. Here there is also the implicit assumption that the weapon system can be 'switched off', as is not the case with biological agents released by a bioweapon.

   9   Ronald Arkin (2010) has argued in favour of the use of such weapons.

   10   It is not a targeted killing.

   11   Further, the villain is legally and morally responsible for foreseeable but unintended killing done by the killer-dogs in the forest, if they had happened upon one of the birdwatchers well known to frequent the forest and mistakenly killed him instead of the intended victim. (Perhaps the birdwatcher carried the scent of birds often attacked by the killer-dogs.)

   12   For criticisms see Steinhoff (2013).

   13   Moreover, each is fully morally responsible; not all cases of collective moral responsibility involve a distribution of the quantum (so to speak) of responsibility. See Miller (2006).

human programmer-activator. Surely this human programmer-activator now has *full* individual moral responsibility for the killing, as the villain does in (both versions of) our killer-dog scenario. To be sure there is no human intervention in the causal process after programming-activation. But the weapon has not been magically transformed from an entity only with causal responsibility to one which now has moral or other than causal responsibility for the killing.

It might be argued that the analogy does not work because killer-dogs are unlike killer-robots in the relevant respects. Certainly dogs are minded creatures whereas computers are not; dogs have some degree of consciousness and can experience, for example, pain. However, this difference would not favour ascribing moral responsibility to computers rather than dogs; rather, if anything, the reverse is true. Clearly, computers do not have consciousness, cannot experience pain or pleasure, do not care about anyone or anything (including themselves) and cannot recognise moral properties, such as courage, moral innocence, moral responsibility, sympathy or justice. Therefore, they cannot act *for the sake of* moral ends or principles *understood as moral in character*, such as the principle of discrimination. Given the non-reducibility of moral concepts and properties to non-moral ones and, specifically, physical ones,[14] at best computers can be programmed to comply with some *non-moral proxy* for moral requirements. For example, 'Do not intentionally kill morally innocent human beings' might be rendered as 'Do not fire at bipeds if they are not carrying a weapon or they are not wearing a uniform of the following description'. I return to this issue below.

Notwithstanding the above, some have insisted that robots are minded agents; after all, it is argued, they can detect and respond to features of their environment and in many cases they have impressive storage/retrieval and calculative capacities. However, this argument relies essentially on two moves that should be resisted and are, in any case, highly controversial. Firstly, rational human thought, notably rational decisions and judgements, are down-graded to the status of mere causally connected states or causal roles, for example via functionalist theories of mental states. Secondly, and simultaneously, the workings of computers are upgraded to the status of mental states, for example via the same functionalist theories of mental states. For reasons of space I cannot here pursue this issue further. Rather I simply note that this simultaneous down-grade/upgrade faces prodigious problems when it comes to the ascription of (even non-moral) autonomous agency. For one thing, autonomous agency involves the capacity for non-algorithmic inferential thinking, for example the generation of novel ideas. For another, computers do not have interests or desires, do not pursue end-in-themselves and cannot choose their own ends. At best they can select between different means to the ends programmed into them. Accordingly, they are not autonomous agents, even non-moral ones. For this reason alone, robopocalypse is evidently an illusion; robotic weapons are morally problematic, but not for the reason that they are autonomous agents in their own right.

---

14   The physical properties in question would not only be detectable in the environment but also be able to be subjected to various formal processes of quantification and so on.

Granted that so-called 'autonomous' human out-of-the-loop weapons are not autonomous (morally or otherwise), nevertheless it has been argued that there is no in principle reason why they should not be used. (Moreover, they are held to have certain advantages over human in-the-loop and on-the-loop systems, for example being machines they are not subject to psychological fear and associated stress (Arkin 2010).) A key claim on which this argument is based is that moral principles, such as military necessity, proportionality and discrimination, can be reduced to rules, and these rules can be programmed in to computers (Arkin 2010).[15] However, I suggest that the phenomenon of moral ramification presents a critical, if not insurmountable, problem at this point. To recap this phenomenon: the combination of logical interdependence between the three *jus in bello* principles and their applicability at all interconnected individual and collective levels gives rise to *moral ramification* and the associated need for complex decision-making, such that one cannot simply apply one of these principles in a given conceptually discrete and self-contained context involving the use of lethal force without taking into account the other principles and other contexts at other levels.

Let us revisit what this might mean in practice. Appropriate applications of, say, the principle of military necessity involves reasonably reliable, morally informed, contextually dependent judgements at the various collective levels, as well as at the individual level, and at the various centres of individual and collective responsibility. However, given the nested character of the individual and collective ends in play, their necessarily underspecified content and the need to be responsive to the actions, including counter-measures, of enemy combatants and their leaders, there is a constant interplay between the various collective and individual levels (for example strategic commanders at headquarters and combatants in a firefight), and across centres (for example different theatres of war). Further, the various applications of the principles of necessity, proportionality and discrimination are logically interdependent, for example the application of the principle of proportionality depends on considerations of military necessity and vice-versa. Accordingly, there is a need to adjudicate not only between the means to given ends, but also with respect to the moral weight to be accorded different competing ends at different levels. For example, the individual end to advance to assist a comrade-in-arms coming under heavy fire might compete with the collective end of one's platoon or company to make a tactical retreat to avoid heavy losses. Again, the collective military end to win firefights might be facilitated by relatively permissive ROE, but perhaps this end competes with the collective end to avoid large-scale casualties among civilians, and the latter

---

15 This claim has been countered by various critics, however not, in my view, decisively. For these critics have, as far as I am aware, relied on piecemeal objections (so to speak), such as the difficulty an autonomous weapon would have in distinguishing innocent civilians from terrorists in civilian dress. See, for example, Sharkey (2012, pp. 111–28). However, a more decisive, and by contrast, *holistic* objection can be made to the application of these principles: the moral ramification argument.

end is facilitated by relatively restrictive ROE. Further, at the macro-collective level, the collective end of the military leadership to win an internecine war might compete with the collective end of the political leadership not to inflict losses of a magnitude that would undermine the prospects for a sustainable peace.

In the light of this let us see what this implies for the project to reduce the three *jus in bello* principles to rules and program them into armed robots. First, each *moral* principle needs to be expressible in a sharply defined rule couched in *non-moral* descriptive terms. Given the non-reducibility of the moral to the non-moral (physical?), it is extremely doubtful that this can be done for moral principles, especially ones that are relatively vague and quite general in form, as are the ones in question. Moreover, even if it could be done, the principles are *logically interdependent* and this would need somehow to be accommodated; logically independent rule specifications, for example, would not work. Second, many, if not most of the uses of lethal force in question are *joint* actions and joint actions are not reducible to aggregations of individual actions (see Miller 1992). So the rules in question would need somehow to accommodate this; the mere aggregation of instructions for single actors, for example, would not suffice. Third, the sharply defined rules in question would presumably be applicable to sharply defined, discrete, self-contained contexts involving the use of lethal force; otherwise the robot would not be able to comply with them. Here the phenomenon of moral ramification comes fully into its own. For, as our above examples demonstrated, in any such conceptually discrete and self-contained context, be it a one-against-one encounter, a firefight, an air strike or a battle, there will inevitably be moral considerations emanating from some other context (for example, another battle) or some larger context of which the discrete, self-contained context is an element (for example, the war as a whole), which bear upon it in a manner that morally overrides or qualifies compliance with the sharply defined rule in question (or set of rules, for that matter[16]). Given that each war taken in its totality is unique, this interplay of contexts has the effect of making decisions in accordance with the *jus in bello radically contextually dependent* and, as such, beyond the reach of sharply defined rules. I conclude that this 'computerised' conception of the application of fundamental moral principles in war faces prodigious, if not insurmountable, problems. In short, evidently robopocalypse is doubly an illusion.

An important consequence of this is that the design, construction and use of human *out*-of-the-loop weapons are highly morally problematic. Such weapons cannot be programmed to comply with the moral principles of military necessity, discrimination and proportionality. Moreover, their use would seriously impede the capacity of their human operators to adequately comply with these moral principles and, to this extent, it would be an abnegation of moral responsibility on the part of the military. Finally, the use of these human *out*-of-the-loop weapons is evidently

---

16   The sharply defined computerised rule conception could be complicated by adding meta-rules, for example. However, this would not make any material difference to the problems; it would simply elevate things to a higher level of complexity.

unnecessary since, as we saw above, for the combat situations in which human-*in*-the-loop weapons are inadequate, human *on*-the-loop weapons are available. I conclude that human out-of-the-loop weapons morally ought not to be used.

References

Arkin, R. 2010, 'The case for ethical autonomy in unmanned systems', *Journal of Military Ethics*, 9: 332–41.

Dennett, D. 1987, *The Intentional Stance*. Cambridge, MA: Bradford Books.

Miller, S. 1992, 'Joint action', *Philosophical Papers*, 21: 275–99.

Miller, S. 1995, 'Intentions, ends and joint action', *Philosophical Papers*, 24: 51–67.

Miller, S. 2006, 'Collective moral responsibility: An individualist account', *Midwest Studies in Philosophy*, 30: 176–93.

Miller, S. 2007, 'Against the moral autonomy thesis', *Journal of Social Philosophy*, 38: 389–409.

Miller, S. 2009, *Terrorism and Counter-terrorism: Ethics and Liberal Democracy*. Oxford: Blackwell Publishing.

Miller, S. 2010, *The Moral Foundations of Social Institutions: A Philosophical Study*. New York: Cambridge University Press.

Miller, S. 2014, 'The Global Financial Crisis and Collective Moral Responsibility', in (ed.) Andre Nollkaemper, *Distribution of Responsibilities in International Law*. Cambridge: Cambridge University Press.

Miller, S. and Blackler, J. 2005, *Ethical Issues in Policing*. Aldershot: Ashgate.

Miller, S. and Makela, P. 2005, 'The collectivist approach to collective moral responsibility', *Metaphilosophy*, 36: 634–51.

Sharkey, N. 2012, 'Killing made easy: From joysticks to politics', in P. Lin, K. Abney and G.A. Bekey (eds), *Robot Ethics: The Ethical and Social Implications of Robotics*. Cambridge, MA: MIT Press.

Sparrow, R. 2007, 'Killer robots', *Journal of Applied Philosophy*, 24, 63–77.

Steinhoff, U. 2013, 'Killing them safely: Extreme asymmetry and its discontents', in B.J. Strawser (ed.), *Killing by Remote Control: The Ethics of an Unmanned Military*. Oxford: Oxford University Press.

Uniacke, S. 2011, 'Proportionality and self-defense', *Law and Philosophy*, 30: 253–72.

Walzer, M. 1977, *Just and Unjust Wars: A Moral Argument with Historical Examples*. New York: Basic Books.

Chapter 13

Singularity and the Art of Warfighters: The Geneva Convention on Trial

Joseph Savirimuthu

This is his fifth upgrade. Corporal Marcus, a soldier from the Elite Shock Regiment taps on the BrainInterface app on his communication device to upgrade a computer chip implanted in his head. The chip possesses a number of functionalities, including the provision of biofeedback and erasure of traumatic battlefield experiences. Early this afternoon, the entire regiment were given intravenous drugs designed to suppress fear and anxiety and help generate cells to self-heal damaged tissues and organs. Recent intelligence suggests that the enemy is likely to use a number of well-known chemical and biological weapons.

Technological and biological enhancement of warfighters may be a game changer. The expectation here is that convergence in technology and sciences will help alleviate many natural human frailties of warfighters. Whether one accepts Ray Kurzweil's optimistic view of our post-human futures and singularity, modern warfare is already being transformed by innovation and information communication technologies. The potential use of new technological and biological resources to augment warfighters with enhanced biological and technological capabilities may end up adding to growing unease about the speed with which new technologies are de-humanising warfare. Do we need to anticipate these new developments by putting in place technology-specific regulations? How should organisations such as the International Committee of the Red Cross (ICRC) or International Committee for Robot Arms Control (IRAC) respond to futuristic warfighters in the rapidly evolving battlefield? In their comprehensive study of enhanced warfighters, Lin et al. (2013, p. 29) conclude that regulatory oversight could be maintained under Geneva Additional Protocol I (API) Article 36 by regarding enhanced warfighters as new weapons. I suggest that understanding the nature of the problem of disconnection is an essential prelude to ensuring that debates on emerging and futuristic technologies reflect the complexity of interactions between law, technology and policy.

Human Enhancement and Singularity

Human enhancement is not a new or novel concept (British Medical Association, 1909). Technological augmentation, digital devices and medicine are available to civilians not only for therapeutic purposes but to improve cognitive capabilities, reduce stress and anxiety and improve performances at work or sport (Bostrom 2008). Each advance in science and medicine for therapeutic and other needs have been accompanied by close scrutiny of legal, ethical and social issues (Academy of Medical Sciences et al. 2013). Human enhancement projects are not confined to the civil space. The military too has invested heavily in research and development in helping increase the successful outcome of military engagements and deployment (National Research Council 2012, pp. 28–31). Ready accessibility and affordability of technologies and innovations in the public domain has led to their direct use, has been adapted for specific use or provided a foundation for creating military-specific applications. Innovative use of information communication technologies, robotics and autonomous systems are becoming a common feature of modern warfare. The use of autonomous weapons systems and robotics for military operations can be seen as a partial response to increased threats posed by a diverse range of enemy combatants and weapons such as improvised explosive devices, and sophisticated disruptive and biological agents. Advances in biological sciences have also been harnessed for military applications to ensure that threats posed by chemical and biological weapons can be neutralised (Huston 2010, pp. 112–17). In a report produced by the Office of the USAF Chief Scientist (2010, p. 59), the prevailing view is that the progress of science is unlikely to be curbed; the convergence between technology, biology and science will lead to the augmentation of human performance through:

> implants, drugs or other augmentation approaches to improve memory, alertness, cognition and visual/aural acuity. It may even extend to limited direct brainwave coupling between humans and machines, and screening of individual capacities for key specialty codes via brainwave patterns and genetic correlators. Adversaries may use genetic modification to enhance specific characteristics or abilities. Performance augmentation will find routine use in the cockpit, on the flight line, by ISR operators and by commanders. Data may be fused and delivered to humans in ways that exploit synthetically augmented intuition to achieve needed decision speeds and enhance decision quality. Human senses, reasoning, and physical performance will be augmented using sensors, biotechnology, robotics, and computing power.

An enhanced warfighter, in the context of this study, can be understood in the sense described by Juengst (1998, p. 29) of an individual who has been the subject of technological and medical or biological intervention designed 'to improve human form or functioning beyond what is necessary to sustain or restore good health'.

Military application of enhancement technologies has two determinants – the first is technological and the second, biological. Technological singularity has long been regarded by advocates of singularity as the next phase in the evolution of humanity. Enhanced warfighters may sometimes be mistakenly seen as foreshadowing a vision of humans being proxies for powerful computing capabilities and intelligence. Technological singularity in this context is usually understood in the sense envisaged by the American mathematician and science fiction writer Vernor Vinge, who coined the term itself. In a well-known 1993 essay titled 'The Coming Technological Singularity: How to Survive in the Post-human Era', Vinge hypothesised that should the trend in the exponential development in software capability and hardware resourcing continue, the possibility of machine intelligence rather than humans driving the next paradigm shift cannot be ruled out (a concept known as the event horizon thesis). Critical to these paradigm shifts is the doubling of computing and processing capabilities in information technology. There is a growing conviction that the convergence of hardware and software will lead to artificial intelligence (AI) technologies becoming less dependent on humans for their creation (Chalmers 2010). The singularity project has its origins in research undertaken to develop machines, which replicate thinking processes we associate with humans (Turing 1950; Good 1970; von Neumann 1966). In the future, super-intelligent machines will transcend Deep Blue (the famous chess-playing computer developed by IBM); the latter operates only within a particular domain. There is considerable uncertainty whether super-intelligent AI will in effect involve the use of brain implants (Shulman 2010). Even though enhanced warfighters do not fit neatly into the singularity hypothesis, it is fair to say that this category of soldiers will have their capabilities enhanced sufficiently to help them adapt and solve problems across multiple domains in a way that would not be possible for ordinary humans. If technological singularity offers opportunities for military applications, advances in biotechnology, neuroscience and the emergence of synthetic biology offer the military additional opportunities. Synthetic biology, for example, can be regarded as one development resulting from the convergence in science and engineering (Silver 2006). The Action Group on Erosion, Technology and Concentration (2007) describes this process as involving 'the design and construction of new biological parts, devices, and systems that do not exist in the natural world and also the redesign of existing biological systems to perform specific tasks'. Consequently, synthetic biological building blocks can now be utilised to create organisms and cells to mirror existing biological functions or even enhance or create novel features (Paradise and Fitzpatrick 2012; OECD, 2014). These developments do not take place in a policy vacuum. Governments are under increasing pressure to reduce costs incurred in fighting expensive wars. Additionally, it has become difficult for governments to ignore public sensitivities to soldier and civilian fatalities resulting from use of new technologies in armed conflicts. Harnessing the potential of new technologies is not only seen as attaining these ends but will serve to provide armies with a strategic advantage (Hammes 2010, pp. 6–7).

The Enhanced Warfighters as Weapons Thesis

There is no unlimited right to the use of any method or means of warfare. API Article 35 provides a general normative framework:

> 2. It is prohibited to employ weapons, projectiles and material and methods of warfare of a nature to cause superfluous injury or unnecessary suffering.

> 3. It is prohibited to employ methods or means of warfare which are intended, or may be expected, to cause widespread, long-term and severe damage to the natural environment.

These principles provide an overarching framework for API, Article 36:

> In the study, development, acquisition or adoption of a new weapon, means or method of warfare, a High Contracting Party is under an obligation to determine whether its employment would, in some or all circumstances, be prohibited by this Protocol or by any other rule of international law applicable to the High Contracting Party.

Taking API Article 35(2) as an example, any determination of the legality of a weapon or means or method will require an assessment, for example, of not only the context and manner of use, its purpose and motivations in the design but whether the military advantage obtained as a consequence causes injury or suffering of a disproportionate nature (Fenrick 1990, p. 500). Customary rules and International Law do not cover enhanced warfighters as such (Rappert et al. 2012; Backstrom and Henderson 2012). This has not stopped humanitarian organizations expressing concern that warfighters of the future may possess machine and biological capabilities, which render human agency obsolete. Not far from these concerns are those which are concerned with warfighters being coerced into permitting technological or biological augmentation or worse, the creation by stealth of a class of warfighters who pay no heed to humanitarian values. It is beyond the scope of this chapter to delve into the issue why such a view is untenable. It is these concerns that perhaps lead to the suggestion that enhanced warfighters could be treated as weapons under the API since they can either be viewed as weapons or a means or method. If this holds true, 174 States that are parties to the API will have an obligation to undertake a review of the lawfulness of an enhanced warfighter. Emerging technologies such as enhanced warfighters raise four regulatory dilemmas: first, the determination of the appropriateness of banning or restricting the use of enhanced warfighters; second, uncertainty regarding the application of existing international laws of war to the use of these technologies in a military context; third, the dangers of duplicating or over-extending existing rules; and fourth, time lag entailed in replacing out-dated rules with new rules and norms. Attempts to address the problems of regulating new or disruptive technologies translate frequently into calls for intervention by policymakers or regulatory authorities. For example, innovations in the field of biotechnology prompted calls to address the uncertain consequences of manipulating 'life forms at the genetic level' (Rhodes 2010, p. 22). The ICRC's initiative on *Biology, Weapons and Humanity* ensures that the international community remain vigilant in ensuring that the rules on chemical and biological weapons are strictly observed. Indeed, as a study by National Research Council (2001, p. 63) pointed out, advances in medicine could also be used to optimise the aggressive capabilities of combatants. Lin et al. (2013, p. 30) propose an analogy, which has the effect of connecting technology to regulation:

> If a war-rhino should be subject to Article 36, then so should this radically enhanced human animal, so it would seem. And to avoid the difficult question of drawing the line at which the enhanced human becomes a weapon, a more intuitive position would be that the human animal is a weapon all along, at every point in the spectrum, especially given the previous reasons that are independent of this demarcation problem.

This is a creative use of an analogy to bridge the gap between 'new' and 'old' weapons, means or methods of warfare. Leaving aside the issue of whether this analogy withstands scrutiny, the pre-emptive turn does goes some way towards attempting to grapple with the legal dilemma in filling the perceived regulatory vacuum. However, there is an alternative approach that may help us better reflect the challenges and opportunities posed by futuristic technologies such as enhanced warfighters. If one were attempting to ensure that the right regulatory framework is in place (assuming one was needed), it would be reasonable to identify the nature of the regulatory problem, describe the significance of the problem of disconnection between law and technology and provide responses to the questions raised. In the accompanying discussion, I would like to articulate the challenges and opportunities posed by enhanced warfighters by using the concept of regulatory disconnection (Brownsword 2008, pp. 161–6). The term regulatory disconnection can be understood in the sense of changes ushered in by new technologies. The rapidity of changes may be lead to a destabilisation of settled norms, rules and values. In some cases, innovations may raise health and safety concerns for individuals or the environment. In others, new technologies may be used to harm or seriously injure individuals or moral interests (Beyleveld and Brownsword 2012). The war rhino analogy is a reasonable attempt to ensure that given the uncertain impact of enhanced warfighters, provisions such as API Article 36 can be used to provide regulatory oversight. As I will argue, we do not have to agree with the characterisation of enhanced warfighters as weapons to address what is in essence a form of descriptive rather than legal disconnection.

The Problem of Regulatory Disconnection

Regulatory disconnection represents the state of affairs that characterise public and policy debates. Describing the complications that emerge with new technologies, Moore (2005, p. 115) captures the recurring dilemmas of regulatory disconnection well:

> We need to formulate and justify new policies (laws, rules, and customs) for acting in these new kinds of situations. Sometimes we can anticipate that the use of the technology will have consequences that are clearly undesirable. As much as possible, we need to anticipate these and establish policies that will minimize the deleterious effects of the new technology. At other times the subtlety of the situation may escape us at least initially, and we will find ourselves in a situation of assessing the matter as consequences unfold. Formulating and justifying new policies is made more complex by the fact that the concepts that we bring to a situation involving policy vacuums may not provide a unique understanding of the situation. The situation may have analogies with different and competing traditional situations. We find ourselves in a *conceptual muddle* about which way to understand the matter in order to formulate and justify a policy.

Brownsword's problem of regulatory disconnection helps us re-evaluate assumptions and our understanding of the nature of disconnection and provides us with a foundation for ensuring that solutions being proposed are in response to the right questions being asked. He proposes (p. 166) that when responding to disconnection, three distinctions are made: (i) descriptive and normative disconnection; (ii) productive and unproductive disconnection; (iii) intelligent and unintelligent purposive reconnection. The following summarises key ideas from these three sets of distinctions.

Descriptive and Normative Disconnection

One can think of numerous examples of descriptive disconnection resulting from technology outpacing the objects or events covered by regulation. We are concerned here with the scope or activity targeted by the regulation being effectively overtaken by technological developments. For example, descriptive disconnection may arise when rules which prohibit playing of radios in the library may not seem to cover headphones which have audio functionality. The librarian may have to consider connecting this technology to the existing rule by either formulating a new rule to cover headphones or design a technology-neutral rule, which designates the space as a quiet zone for study. Rapid technological developments, which lead to descriptive disconnection, create opportunities for assessing the continued relevance of values in view of evolving social, cultural and political expectations and attitudes. For example, the values underpinning the rules, which limited the ability of minors to enter into contracts, is now being re-examined due to cultural acceptance of children as consumers. During the past two decades the emergence of peer-to-peer networks has compelled industry and policymakers to require a reassessment of rules on copyright law, in view of the destabilisation of pre-existing norms governing intellectual property rights.

Productive and Unproductive Disconnection

Another relevant distinction that cannot be discounted is productive and unproductive disconnection. To avoid over-regulating, or due to the time lag involved between the emergence of the technology and passing of legislation, a technique frequently adopted is to infer the intention of the lawmakers or the mischief designed to be averted. Of course, where regulation has been drafted in general and non-specific terms, such a technique is readily employed. However, where legislation targets specific technology or activity, there are constraints to extending the rules to emerging technologies or new uses of such technologies. Judges have managed to negotiate these constraints where the literal application of rules would have produced an absurd outcome. In the event that a literal interpretation indicates that there is likely to be a legal disconnection, one proposal would be to distinguish between unproductive and productive disconnection (Brownsword 2012, p. 167). The aim here would be to determine whether there is a need for regulatory intervention. A purposive approach, which coheres with the intention of the legislators, would be an example of unproductive disconnection. In this type of disconnection, we are concerned with matters which are unlikely to be contentious. Productive disconnection by contrast would require policymakers to avoid purposive interpretation where the emerging technologies make it difficult to make proper assessments of risks or even ascertain how the concept of human dignity is now to be understood. Examples of productive disconnection include innovations involving stem cell research, *in vitro* fertilisation and human enhancement. These are examples which involve multiple regulatory domains and concern disagreements about how dual-use research innovations are to be regulated. Imagine that we uncritically transpose API Article 36 to human enhancement in the military context and the resulting policy ramifications in domains such as health, neuroethics (Levy 2007; Moreno 2012) and bioethics (Glannon 2006; Wolfendale 2008).

Intelligent and Unintelligent Disconnection

Brownsword's two sets of distinctions figure prominently when policymakers and society are faced with having to determine how best to respond to technological changes. There is an additional set of distinctions that may help inform decisions on whether regulatory intervention is to be resisted. The intelligent/unintelligent disconnection category is concerned not so much with ensuring that the right type of deliberation and analysis is pursued, but that the correct decision is taken on whether to fill the perceived regulatory void. The *raison d'être* is that sub-optimal regulatory strategies, as Brownsword (2012, p. 167) suggests, may ultimately prove to be counterproductive or compound the difficulties:

> Putting this in terms of intelligent and unintelligent purposive reconnection, interpreters do the intelligent thing if they employ a purposive approach to reconnect in the case of unproductive descriptive disconnection; but they do not act intelligently if they take a purposive approach to reconnect when, in fact, the disconnection (whether descriptive or normative) is productive and invites more general debate.

In the next section, I would like to briefly explain how the regulatory dilemma raised by enhanced warfighters could be better structured to ensure that we can move towards creating an appropriate regulatory environment.

'War Rhinos', Corporal Marcus and Article 36: A Regulatory Disconnection Lens

What does framing the discourse on the interaction between enhanced warfighters and API Article 36 through the regulatory disconnection lens achieve? One answer is that the regulatory disconnection lens enables us to adopt a set of analytical distinctions that direct us not only to the nature of disconnection but help us recognise the need to engage in the right conversations when assessing how best to create the right regulatory environment. It is also timely. The need to address regulatory gaps created by the rapid pace of innovation and technological advances is one of the drivers in current debates on the urgent need to regulate the use of new technologies in modern warfare such as autonomous weapons systems. Integrating technology into discussions on creating the right regulatory environment is already proving to be a slow and convoluted process. It is now more than two years since the Human Rights Watch (HRW) turned media and public attention towards the way new technologies are transforming modern warfare (HRW 2012). The European Parliament expressed concerns in respect of the 'development, production and use of fully autonomous weapons' and risks posed by delegating critical military tasks to machines (European Parliament 2014). The ICRC in its report highlighted the undesirable humanitarian dimensions resulting from the marginalisation of human agency and responsibility by autonomous weapons systems (ICRC 2014). This concern is further underlined in a statement made by the ICRAC at the Convention on Conventional Weapons Meeting of Experts on lethal autonomous weapons systems (ICRAC 2014) that '[t]he combined strengths of humans and computers operating together with the human in charge of targeting decisions makes better military sense and is necessary in order to meet the requirements of international law'. The United Nations recently convened a meeting in keeping with its remit under the Convention on Certain Conventional Weapons (United Nations 2014) to 'discuss the questions related to emerging technologies in the area of lethal autonomous weapons systems … '. These high-level meetings were preceded by not dissimilar debates at the United Nations Human Rights Council (UNHRC) following publication of a report by the UN Special Rapporteur on lethal autonomous robotic weapons (UNHRC 2013). Addressing the problem of regulatory disconnection is a pervasive theme and customised solutions are proving to be elusive (UNCRC 2013, para. 48):

> The nature of robotic development generally makes it a difficult subject of regulation, especially in the area of weapons control. Bright lines are difficult to find. Robotic development is incremental in nature. Furthermore, there is significant continuity between military and non-military technologies. The same robotic platforms can have civilian as well as military applications, and can be deployed for non-lethal purposes (e.g. to defuse improvised explosive devices) or be equipped with lethal capability (i.e. LARs). Moreover, LARs typically have a composite nature and are combinations of underlying technologies with multiple purposes.

The modern battlefield is now replete with rhetoric such as 'drones', 'remote targeting', 'autonomous weapons systems', 'precision targeting' and 'killer robots'. Dual-use research and technologies illustrate the complexity of managing technological change in the private, public and international sphere. Given the sensitivities of national security and the secrecy within which projects are developed and used, there is some concern that failure to enact regulation 'could lead to an arms race, proliferation, and deployment of technology before it is ready to deal with the potential legal challenges' (Human Rights Watch 2014) How can we begin to articulate the risks posed by this class of enhanced warfighters, if at all, into meaningful weapons review standards under API Article 36 (Savulescu and Bostrom 2009)? Sassoli is of course correct in stating that where determination of the *lex lata* is problematic, recourse can be had to both the purpose and object of treaty rules (Sassoli 2011, p. 48). The quest for clear guidance and criteria are likely to be problematic given the complex nature of enhancements and defining boundaries of lawful and unlawful performance enhancement practices and technologies is likely to be slow and complex. Concerns about the legality of enhanced warfighters parallel arguments encountered in respect of the use of autonomous weapons systems. It is critical to the present discussion that we make clear that enhanced warfighters such as Corporal Marcus are regarded as one means through which strategic or military objectives are likely to be realised. It is however fatal to the enhanced warfighter as a weapons argument that the strained construction does not make rational sense when considered alongside Rule 1 of the Air and Missile Warfare Manual's description of weapons as including guns, missiles or other devices capable of causing injury or death to persons or damage to objects. Enhanced warfighters may represent a new approach to military engagement and problematising disconnection may direct us towards assessing whether there is in fact a need for legislative or other regulatory interventions.

It may very well be that if singularity is achieved in the near future we may then have to resort to precautionary modes of reasoning. At this present time, however, it may be prudent to avoid drawing parallels between autonomous robots and enhanced warfighters along the lines proposed in the study (Lin et al. 2013, p. 29):

> If autonomous robots are clearly regulatable weapons, then consider the spectrum of cyborgs – part-human, part-machine – that exists between robots and unenhanced humans. Replacing one body part, say a human knee, with a robotic part starts us on the cybernetic path. And as other body parts are replaced, the organism becomes less human and more robotic.

Finally, one may also question the appropriateness of connecting enhanced warfighters to API Article 36 that attempts to minimise the frequency of particular forms of soldier fatalities in modern warfare – decision-making under conditions of uncertainty and stress. Soldiers with BrainGate technology will benefit from significant improvements in reaction times. Trans-cranial electrical stimulation apps could be used to enhance memory and cognitive capabilities. Developments in synthetic biology may enable damaged cells to self-repair or be used to manipulate genes to moderate threat anxiety situations. It would seem incoherent if not perverse to permit *ex post* treatment but not *ex ante* enhancement which can minimise the risk of injury or harm arising. This is a point that is not lost on policymakers. The Ministry of Defence's Development, Concepts and Doctrine Centre (DCDC) acknowledges that by 2014 the character of conflict will very likely require the military to engage with adversaries in chaotic environments advantageous to indigenous forces (DCDC 2010, p. 20). Enhancement of physical and cognitive capabilities of warfighters could arguably be consistent with API Articles 35 and 57 if it is part of a strategic choice to minimise civilian casualties and prevent unnecessary suffering.

In summary, warfighters who are augmented do not give rise to normative disconnection under API. Neither is there a need to pursue a purposive disconnection approach – to do so would be result in an unintelligent disconnection since the 'human-in-the-loop' dimension is not removed. It is reasonably well known that dual-use research and technologies raise the dilemma of ensuring that innovation and legitimate use is not stifled by unhindered precautionary reasoning. We need to adopt a cautionary stance when assessing the likely impact of enhanced warfighters. Verbeek (2011, p. 3) is right to observe that too often our responses to new technologies have been shaped by fears and underestimate the relationship between technology and society. Consequently, interpreters can be regarded as doing the intelligent thing if they employ a purposive approach to reconnect in the case of unproductive descriptive disconnection; but they do not act intelligently if they take a purposive approach to reconnect when, in fact, the disconnection (whether descriptive or normative) is productive and invites more considered and informed debate. Creating the right regulatory environment can be particularly problematic when innovations and technologies are combined in a way that may lead to outcomes we either wish to avoid or which may end up providing a catalyst for unintended social consequences (Collingridge 1980, pp. 16–17). Enhanced warfighters can be regarded as an illustration of the 'Collingridge dilemma'. Should we accede to a purposive interpretation when there are pre-existing rules and norms that deal with mischief associated with mature technologies or biological and chemical weapons? What are the costs of ignoring established norms that already regulate the behaviour of warfighters? The intelligent/ unintelligent distinction is intended to remind us of the value of being sensitive to the nature of the disconnection and assessing whether appropriate intervention should in fact require a reassessment of how existing norms and rules can be used to articulate commitment to regulatory obligations. An enhanced warfighter who unleashes chemical and biological toxins on the civilian population would fall foul of existing rules. Indeed, existing multilateral conventions contain normative values that extend to all forms of warfighters. None of the foregoing discussion impinges on the general view that cognitive or technological augmentation *per se*, which gives an army an advantage or neutralises advantages possessed by an opposing force, is permitted unless they use weaponry prohibited by LOAC.

Conclusion

Emerging technologies such as autonomous weapons systems already raise questions about the role of the LOAC and customary rules. Anxiety about the dystopian vision of technology will not only lead to shaping the way we think about futuristic technologies such as enhanced warfighters but may heighten pressures on policymakers to intervene through pre-emptive legislation. The image of an out-of-control warfighter that lacks empathy and morality is suggestive of humanity being at the mercy of science (Winner 1977, p. 17). Conventional thinking assumes that existing interpretations can be extended to enhanced warfighters. Instead of using creative constructions and interpretive techniques to reassure us that law can keep pace with technological innovations, we could use the 'regulatory disconnection' lens to create a space for considered examination of both risks and benefits without resorting to apocalyptic scenarios. As the discussion demonstrates, new technologies or innovations do not invariably lead to legal disconnection. Problematizing technological innovations as disconnection is a prelude to ensuring that debates on enhanced warfighters reflect the complexity of interactions between law, technology and policy. Given that enhancement of warfighters, as defined in the study, appear to be very much in the preliminary phase of development (rather than deployment), the challenge for regulatory connection would be to assess how pre-existing rules and norms can be better harnessed to help guide policymakers and address humanitarian and ethical concerns. The LOAC and protocols have never given technology and science a free rein. At present, enhanced warfighters such as Corporal Marcus will only be too aware that LOAC and multilateral conventions already provide default rules and recognise that any departures must be justified. Problematizing technology through the lens of regulatory disconnection will help us remain vigilant.

References

Academy of Medical Sciences, the British Academy, the Royal Academy of Engineering, and the Royal Society 2013, *Human Enhancement and the Future of Work* viewed 28 June 2014, http://royalsociety.org/uploadedFiles/Royal_Society_Content/policy/projects/human-enhancement/2012–11–06-Human-enhancement.pdf.

Action Group on Erosion, Technology and Concentration 2007, Extreme Genetic Engineering: An Introduction to Synthetic Biology, ETC., viewed 25 February 2014, http://www.etcgroup.org/content/extreme-genetic-engineering-introduction-synthetic-biology.

Backstrom, A. and Henderson, I. 2012, 'New capabilities in warfare: an overview of contemporary technological developments and the associated legal and engineering issues in Article 36 weapons reviews', *IRRC*, 94(886), 483–514.

Beyleveld, D. and Brownsword, R. 2012, 'Emerging Technologies, Extreme Uncertainty, and Principle of Rational Precautionary Reasoning', *Law, Innovation and Technology*, 4(1): 35–65.

Bostrom, N. 2008, 'Drugs can be used to treat more than disease', *Nature*, 451(7178): 520.

British Medical Association 1909, *Secret Remedies. What They Cost and What They Contain* London: British Medical Association.

Brownsword, R. 2008, *Rights, Regulation and the Technological Revolution*. Oxford: Oxford University Press.

Chalmers, D. 2010, 'The Singularity: A Philosophical Analysis', *Journal of Consciousness Studies*, 17(7): 7–65.

Collingridge, D. 1980, *The Social Control of Technology*. London: Pinter.

European Parliament 2014, *European Parliament Resolution of 27 February 2014 on the Use of Armed Drones* (2014/2567(RSP)), viewed 20 July 2014, http://www.europarl.europa.eu/sides/getDoc.do?type=TA&reference=P7-TA-2014–0172&language=EN&ring=P7-RC-2014–0201.

Fenrick, W. 1990, 'The conventional weapons convention: A modest but useful treaty', *IRRC*, 279, 498–509.

Glannon, W. 2006, *Bioethics and the Brain*. Oxford: Oxford University Press.

Good, I. 1970, 'Some Future Social Repercussions of Computers', *International Journal of Environmental Studies*, 1(1–4): 67–79.

Hammes, T. 2010, 'Biotech Impact on the Warfighter', in R.E. Armstrong (ed.), *Bio-Inspired Innovation and National Security*. Washington, DC: National Defense University Press.

Human Rights Watch 2012, *Losing Humanity: The Case Against Killer Robots*, viewed 12 April 2014, http://www.hrw.org/reports/2012/11/19/losing-humanity-0.

Huston, A. 2010, 'Bioenzymes and Defence', in R.E. Armstrong (ed.), *Bio-Inspired Innovation and National Security*. Washington, DC: National Defence University Press.

International Committee of the Red Cross 2014, *Report of the ICRC Meeting on Autonomous Weapon Systems*, viewed 25 July 2014, http://www.icrc.org/eng/resources/documents/report/05-13-autonomous-weapons-report.htm.

International Committee for Robots Arms Control 2014, *CCW Statement by the International Committee for Robots Arms Control*, viewed 25 July 2014, http://www.unog.ch/80256EDD006B8954/(httpAssets)/AED53FFA53455DBBC1257CDA005654A6/$file/NGOICRAC_LAWS_FinalStatement_2014.pdf.

Juengst, E. 1998, 'The Meaning of Enhancement', in E. Parens (ed.), *Enhancing Human Traits: Ethical and Social Implications*. Washington, DC: Georgetown University Press.

Levy, N. 2007, *Neuroethics: Challenges for the 21st Century*. Cambridge: Cambridge University Press.

Lin, P. Mehlman, M. and Abney, K. 2013, 'Enhanced Warfighters: Risks, Ethics, and Policy', Case Research Paper Series in Legal Studies, 1 January 2013, viewed 15 March 2013, http://ethics.calpoly.edu/Greenwall_report.pdf.

Ministry of Defence's Development, Concepts and Doctrine Centre (DCDC) 2010, *Future Character of Conflict*, viewed 20 June 2014, https://www.gov.uk/government/publications/future-character-of-conflict.

Moore, J. 2005, 'Why we need better ethics for emerging technologies', *Ethics and Information Technology*, 7(3): 111–19.

Moreno, J. 2012, *Mind Wars: Brain Science and the Military in the 21st Century*. New York: Bellevue Literary Press.

National Research Council 2012, *Human Performance Modification: Review of Worldwide Research with a View to the Future*. Washington, DC: National Academies Press.

National Research Council 2001, *Opportunities in Biotechnology for Future Army Applications*. Washington, DC: National Academy Press.

Organization for Economic Cooperation and Development 2014, *Emerging Policy Issues in Synthetic Biology*, OECD, Geneva, viewed 14 July 2014, http: 10.1787/9789264208421-en.

Paradise, J. and Fitzpatrick, E. 2012, 'Synthetic Biology: Does Re-Writing Nature Require Re-Writing Regulation?' *Penn State Law Review*, 117: 53–88.

Rappert, B. Moyes, R. Crowe, A. and Nash, T. 2012, 'The roles of civil society in the development of standards around new weapons and other technologies of warfare', *IRRC*, 94(886): 765–85.

Rhodes, C. 2010, *International Governance of Biotechnology: Needs, Problems and Potential*. London: Bloomsbury Academic.

Sassoli, M. 2011, 'The role of human rights and international humanitarian law in new types of armed conflicts', in O. Ben-Naftali (ed.) *International Humanitarian Law and International Human Rights Law*. Oxford: Oxford University Press.

Savulescu, J. and Bostrom, N. 2009, *Human Enhancement*. New York: Oxford University Press.

Shulman, C. 2010, 'Whole Brain Emulation and the Evolution of Superorganism', viewed 20 May 2014, http://intelligence.org/files/WBE-Superorgs.pdf.

Silver, L. 2006, *Challenging Nature*. London: HarperCollins.

Schmitt, M. 2012, 'Classification in future conflict', in E. Wilmshurst (ed.), *International Law and the Classification of Conflicts*. Oxford: Oxford University Press.

Turing, A. 1950, 'Computing Machinery and Intelligence', *Mind*, 59(236): 433–60.

United Nations 2014, viewed 25 July 2014, http://www.unog.ch/80256EE6005 85943/%28httpPages%29/6CE049BE22EC75A2C1257C8D00513E26?Open Document.

United Nations Human Rights Council 2013, *Report of the Special Rapporteur on Extrajudicial, Summary or Arbitrary Executions, Christof Heyns*, A/HRC/23/47. Geneva: United Nations.

Verbeek, P.P. 2011, *Moralizing Technology: Understanding and Designing the Morality of Things*. Chicago: University of Chicago Press.

Vinge, V. 1993, 'The Coming Technological Singularity: How to Survive in The Post-Human Era', in G.A. Landis (ed.), *Proceedings of Vision 21: Interdisciplinary Science and Engineering in the Era of Cyberspace*. Lewis Research Center, NASA, 11–22.

von Neumann, J. 1966, *Theory of Self-Reproducing Automata*, edited and completed by A. Burks. Urbana, IL: University of Illinois Press.

Winner, L. 1977, *Autonomous Technology: Technics-out-of-Control as a Theme in Political Thought*. Cambridge, MA: Massachusetts Institute of Technology.

Wolfendale, J. 2008, 'Performance-enhancing technologies and moral responsibility in the military', *American Journal of Bioethics*, 8(2): 28–38.

Index